Lecture Notes in Artificial Intelligence 2293

Subseries of Lecture Notes in Computer Science
Edited by J. G. Carbonell and J. Siekmann

Lecture Notes in Computer Science

Edited by G. Goos, J. Hartmanis, and J. van Leeuwen

Springer

Berlin
Heidelberg
New York
Barcelona
Hong Kong
London
Milan
Paris
Tokyo

Jochen Renz

Qualitative Spatial Reasoning with Topological Information

 Springer

Series Editors

Jaime G. Carbonell, Carnegie Mellon University, Pittsburgh, PA, USA
Jörg Siekmann, University of Saarland, Saarbrücken, Germany

Author

Jochen Renz
Vienna University of Technology, Dept. of Computer Science, Database and AI Group
Favoritenstr. 9, 1040 Wien, Austria
E-mail: renz@dbai.tuwien.ac.at

Cataloging-in-Publication Data applied for

Die Deutsche Bibliothek - CIP-Einheitsaufnahme

Renz, Jochen:
Qualitative spatial reasoning with topological information / Jochen Renz. -
Berlin ; Heidelberg ; New York ; Barcelona ; Hong Kong ; London ; Milan ;
Paris ; Tokyo : Springer, 2002
   (Lecture notes in computer science ; Vol. 2293 : Lecture notes in
   artificial intelligence)
   ISBN 3-540-43346-5

CR Subject Classification (1998): I.2.9, I.2, H.2.8, I.4

ISSN 0302-9743
ISBN 3-540-43346-5 Springer-Verlag Berlin Heidelberg New York

Springer-Verlag Berlin Heidelberg New York
a member of BertelsmannSpringer Science+Business Media GmbH

http://www.springer.de

© Springer-Verlag Berlin Heidelberg 2002
Printed in Germany

Typesetting: Camera-ready by author, data conversion by PTP-Berlin, Stefan Sossna
Printed on acid-free paper   SPIN: 10846393   06/3142   5 4 3 2 1 0

Preface

Spatial knowledge representation and reasoning with spatial knowledge are important in many application areas such as robotics, geographical information systems, and computer vision. While most research addresses the problems connected with spatial knowledge from a purely quantitative point of view, more recently methods for representing and reasoning with spatial knowledge in qualitative ways have been developed. The motivation behind such qualitative approaches is the desire to represent and reason on a more abstract level that is closer to the way humans think and speak.

Jochen started research in this area when he did his Master's thesis in 1995 under my supervision. In 1996, the German national science foundation DFG initiated a priority research program on spatial cognition, which had as one of its topics research on the computational properties of qualitative, spatial representation approaches. Of course, Jochen became one of the researchers in this priority program. And as this book proves, this was not a mistake! His work in the priority program led to the solution of a number of open research problems and to the contribution of new research methods.

Jochen started out with the qualitative, topological constraint calculus RCC8 proposed by Randell, Cui, and Cohn. This calculus has eight different basic relations and allows one to describe spatial configurations of regions in a very abstract, qualitative way. As had been shown by Bennett – using results by Tarski – reasoning in the calculus is decidable. However, it was by no means clear, whether there are efficient ways to reason in the calculus or fragments of it. It was not clear how to represent the notion of "region" inside a computer, and it was not obvious which role the dimension of space would play for the reasoning process. Even when having answered all these questions, one might wonder what the "practical" performance of reasoning algorithms for RCC8 is? Can we deal with 10, 100, or 1000 regions? Finally, one might wonder whether the RCC8 relations are "cognitively plausible."

Jochen found answers to all of these questions. This implies that he had to use methods from logic and theoretical computer science in order to address questions concerning computational properties. The questions concerning practical performance had to be answered by large-scale empirical computational experiments. Finally, the cognitive questions were answered using the experimental, psychological paradigm. The most impressive result is probably the complete classification of computational properties for all fragments of RCC8. This result was achieved using a new method for determining the computational complexity of constraint calculi with infinite domains, a method that can be used for any qualitative constraint calculus.

All in all, I believe that this book will become one of the standard references in the area of qualitative spatial reasoning. Although it could not solve all problems in this area – which is probably impossible – it is a very large step ahead.

January 2002   Bernhard Nebel

Acknowledgements

First and foremost I would like to thank Bernhard Nebel for his excellent support. It has been a great pleasure working with him and within his group. Very special thanks to Tony Cohn for the time he spent as the external reviewer of my thesis, for the fruitful discussions, and for plenty of good advice. For the numerous useful discussions on the topic of this book and for their helpful comments I would like to thank Brandon Bennett, Yannis Dimopoulos, Christoph Dornheim, Alfonso Gerevini, Steffen Gutmann, Markus Knauff, Gerard Ligozat, Reinhold Rauh, Luca Viganò, and the anonymous reviewers of the papers resulting from this book. I would like to thank Ronny Fehling and Thilo Weigel for their skillful assistance in designing and implementing experiments and algorithms. Finally, I would like to thank the Deutsche Forschungsgemeinschaft which supported this work under grants Ne 623/1-1 and Ne 623/1-2 as part of the DFG special research effort on *Spatial Cognition* as well as under the doctoral program on *Human and Machine Intelligence* at the Albert-Ludwigs University Freiburg. Both institutions provided me with an excellent interdisciplinary scientific environment.

Remarks

This book is a slightly revised version of my PhD thesis submitted to the Faculty of Applied Science at the Albert-Ludwigs-University of Freiburg, Germany in May 2000. Parts of it have already been published in journals, in conference proceedings, and in collections. Parts of Chapter 5 appear in the *Proceedings of the 3rd International Conference on Spatial Information Theory (COSIT'97)* [102], other parts have been published in the collection *Spatial Cognition II* [145]. The psychological background (Section 5.1.) is based on a first version written by Reinhold Rauh. Parts of Chapter 6 have been published in the *Artificial Intelligence* journal [143], in the *Proceedings of the 15th International Joint Conference on Artificial Intelligence (IJCAI'97)* [141], and in the *Proceedings of the 4th International Conference on Principles and Practice of Constraint Programming (CP'98)* [68]. Parts of Chapter 7 appear in the *Proceedings of the 16th International Joint Conference on Artificial Intelligence (IJCAI'99)* [140], parts of Chapter 8 in the *Journal of Artificial Intelligence Research* [144] and in the *Proceedings of the 13th European Conference on Artificial Intelligence (ECAI'98)* [142], and parts of Chapter 9 in the *Proceedings of the 6th International Conference on Principles of Knowledge Representation and Reasoning (KR'98)* [139].

Table of Contents

List of Tables

List of Figures

1. Introduction

Space is a fundamental part of our life. Our whole existence is embedded in space. All physical entities are located in space, and throughout our life we change location, shape, and other spatial properties of the world around us. In order to do so deliberately and intelligently, we must be able to deal with spatial knowledge effectively. We have to know where things are, how to get there, and how to change things in order to reach our goals. Our spatial knowledge is obtained and modified by what we perceive from the external world and by our ability to derive new knowledge. We must also be able to communicate spatial knowledge with others.

Certainly, some of the major abilities and requirements of any intelligent system are to perceive space, to represent space, to reason about space, and to communicate about space. Therefore, it is an important task of Artificial Intelligence research to provide means for dealing with spatial knowledge.

1.1 Different Approaches for Representing Spatial Knowledge

The basic task is to find an adequate formalism for representing spatial knowledge. All other tasks such as the reasoning mechanisms depend on the representation formalism. There are mainly two different approaches for representing spatial knowledge, the quantitative approach and the qualitative approach. The quantitative, numerical approach is the classical approach which is used in many traditional disciplines like computer vision [98], computational geometry [133] and robotics [113]. In this approach it is common to have a (global or local) coordinate system in which all spatial information is maintained. A possible implementation of this approach would be to have several agents each having a local coordinate system that all contribute in building or modifying a global coordinate system. Spatial knowledge can be communicated, for instance, by mapping the local with the global coordinate system and exchanging the relevant global coordinates. In this case, all participating agents must have access to the same global coordinate system (cf. [80]). In the quantitative approach the exact position and extent of all objects should be known as it is difficult to deal with indefinite or inexact

J. Renz: Qualitative Spatial Reasoning with Topological Information, LNAI 2293, pp. 1–11, 2002.
© Springer-Verlag Berlin Heidelberg 2002

knowledge. It is also difficult to deal with rules that can only be applied if certain spatial configurations occur. Reasoning within this approach is usually done with numerical or geometrical methods like computing a tangent to an object or computing the closest point of one object to another.

The qualitative approach is a novel alternative for representing spatial knowledge without using exact numerical values [165, 34]. Instead, spatial information is represented using a finite vocabulary such as a finite number of possible relationships. *Qualitative spatial representations* [27, 87] appear to be closer to how humans represent and communicate spatial knowledge, namely, by specifying the relationships of and between spatial entities. Most spatial descriptions in natural language such as "A is above B", "A is between D and E", "A is larger than B", "C is inside D", "B is next to D", or "C touches E" are of this kind. Spatial descriptions like these are usually sufficient for us to identify a described object or follow a described route. Using the qualitative approach it is immediate to represent indefinite knowledge such as "C touches E from the left or from the right". It is also easily possible to specify and apply rules like "if part P arrives in L, put P into the box B and bring B to A". Although exact numerical values are not used, qualitative spatial knowledge is not inexact. Distinctions are only made if necessary and depend on the chosen level of granularity, i.e., the chosen set of different relationships used to describe a spatial situation. For instance, given a task such as "Get the book from the desk and put it in the left drawer", the exact position of the book or the drawer is not necessarily important. Unless there is more than one book on the desk or more than one left drawer, the formulation of the task is sufficient. Reasoning within this approach, *qualitative spatial reasoning*, can be done by deriving spatial relationships from the given knowledge. Given, for instance, the following spatial descriptions "I left my wallet in the car" and "the car is in the garage", we can derive that the wallet is also in the garage.

Both, the coordinate based and the qualitative approach have their own right. For some applications the coordinate based approach is better suited, for others the qualitative approach. Whenever human interaction is involved, the qualitative approach seems to be the better choice. Nevertheless, if exact coordinate based knowledge is available it should be used. As an example, consider a navigation task. When trying to find a way in an unknown city, a route description given by another person is usually purely qualitative, e.g., "go straight, move to the right at the second crossing, move to the left after passing the church...". A street map which represents coordinate based knowledge can be used to extract the same qualitative description. A car navigation system may combine a coordinate based map with a GPS system in order to obtain a qualitative description, e.g., "turn to the right at the next crossing". Thus, from an applicational perspective, both kinds of representations are important and it seems to be interesting to have both kinds in parallel and the means to transform one representation into the other one and *vice-versa*.

1.2 Qualitative Spatial Representation and Reasoning

In this work we focus on the qualitative approach, i.e., spatial knowledge is represented by specifying qualitative relationships of and between spatial entities. As indicated by the natural language expressions above, there are different kinds of spatial relationships representing different aspects of space such as orientation, distance, size, topology, or shape. Although the different aspects are not independent of each other, for instance, if A is larger than B, then A cannot be contained in B or if C is far from D, then C cannot touch D, most existing approaches to qualitative spatial representation take only one aspect into account. When developing a qualitative formalism for one of these aspects, it is meanwhile common practice to try to identify a suitable set of jointly exhaustive and pairwise disjoint relations called *base relations* that cover the desired aspect on a certain level of granularity.

Using base relations for representing spatial knowledge has several advantages. Between any n spatial entities exactly one n-ary base relation holds (Note that in this work we are only dealing with binary relations.) Thus, indefinite knowledge can be expressed as a union of different possible base relations. A further advantage of using base relations for representing spatial knowledge is the possibility of applying constraint-based reasoning mechanisms originally developed for qualitative temporal representation and reasoning [110] (which can be regarded as a special case of qualitative spatial representation and reasoning since in the usual understanding[1] time is viewed as a one-dimensional oriented space). In order to apply these methods, spatial knowledge is expressed in the form of constraints xRy where x, y are variables for spatial entities and R is a base relation or a union of different base relations. Most spatial reasoning tasks such as inferring new constraints from a set of given constraints or checking whether a certain constraint holds, can then be reduced to computing consistency of a set of constraints [69, 161], i.e., checking whether it is possible to assign all variables with spatial entities such that all constraints are satisfied. For this form of reasoning it is necessary to have a composition table that contains all possible compositions of base relations, i.e., for every pair of base relations R_1, R_2 the composition table contains the relation R_3 which is the union of all possible base relations that can hold between x and z if xR_1y and yR_2z hold for arbitrary spatial entities x, y, and z. Consistency can be approximated by enforcing path-consistency to the set of constraints, which corresponds to making all triples of constraints consistent.

---

[1] In qualitative temporal and spatial representation and reasoning we are not concerned with relativistic aspects of time and space, but with a "naive" view of time and space.

Having such standard methods for qualitative spatial reasoning is very favorable. What is needed is basically a set of base relations and the corresponding composition table. However, the compositions should be computed using the formal semantics of the relations. Otherwise it is not possible to prove soundness and completeness of the reasoning mechanisms. For some approaches to qualitative spatial reasoning no formal semantics of the relations were given and it appears that in these cases the composition tables were obtained manually based on the intuitive meaning of the relations. This is not only impossible for large sets of base relations, there is also no way of formally verifying the given compositions and, therefore, of proving tractability or even decidability of the reasoning mechanisms.

In comparison with qualitative spatial reasoning, it is common in qualitative temporal reasoning to give formal semantics of the relations and to prove decidability and complexity of the reasoning problems (e.g., [4, 128, 115, 110, 160]). However, providing temporal relations with formal semantics is much easier than it is for spatial relations. Temporal entities are either time points or time intervals which can be represented by their endpoints. Thus, temporal reasoning is mainly reasoning about ordered points, no matter what aspects of temporal entities are taken into account, order, distance, or extension. In this case, the formal semantics is obvious—for instance it is possible to use the real numbers. The same holds for one-dimensional space, but for spatial entities in two- or higher dimensional space it is not at all clear how they can be represented and, therefore, how the spatial relationships can be provided with formal semantics. Some approaches to qualitative spatial representation and reasoning try to adopt the results and the formal semantics of temporal (one-dimensional) reasoning by projecting entities of n-dimensional space to the defining axes and apply the one-dimensional relationships to the single axes [79, 6]. This works only if all spatial entities are cuboids whose sides are parallel to the axes, otherwise all represented relationships are approximations to the actual relationships.

It is one of the main motivations for qualitative spatial representation and reasoning that it is considered to be akin to human spatial cognition [57]. Thus, apart from the formal properties, it should be an important criterion for choosing a set of spatial relations that humans use the same relations for distinguishing spatial configurations. This is also referred to as (conceptual) cognitive adequacy [153]. If this criterion is not taken into account, it is not immediate to communicate the represented spatial knowledge with humans. It is often claimed in the literature that some approach is cognitively adequate or cognitively valid. This claim, however, must be based on empirical justification and not only on the researchers' intuition.

1.3 Applications and Research Goals of Qualitative Spatial Representation and Reasoning

Since dealing with spatial knowledge is—either explicitly or implicitly— necessary for any intelligent system, there are many potential though few existing applications of qualitative spatial representation and reasoning. Some possible and existing applications include spatial information systems [81, 14, 47], configuration tasks, (robot) navigation [117, 49], computer vision, natural language processing [117], document analysis [163], visual (programming) languages [70], and qualitative simulation of physical processes [33, 136, 108]. One reason for the lack of existing applications is that most applications require more than one aspect of space. Representing and reasoning about, e.g., orientation or size only is most often not sufficient. Since the different aspects are not independent of each other, it is not possible just to add different approaches covering the single aspects in order to cover all necessary aspects. Although this is supported by the constraint-based reasoning mechanisms, it is necessary to have an additional module that controls the possible interconnections of the different constraints. Another reason for the lack of existing applications is that for most approaches there are no efficient and sometimes even no complete reasoning mechanisms. These, however, are essential for most applications.

Hence, the goal of research in qualitative spatial representation and reasoning must be to develop powerful representation formalisms which cover many aspects of space, which are provided with formal semantics, and which allow for efficient reasoning. Furthermore, since similarity to human spatial cognition is an important motivation for qualitative spatial representation and reasoning, the developed approach should also be cognitively adequate.

1.4 Topological Relations as a Basis for Qualitative Spatial Representation and Reasoning

In this work we establish a basis for reaching this goal. We are concerned with one of the most fundamental aspects of space, namely, its topology. Topological relationships are those relationships that are invariant with respect to continuous transformations of the underlying space: "C is inside D", "B overlaps D", or "C touches E" are examples for topological relationships. In Cartography, for instance, there are different ways for designing geographical maps of the world. Some maps preserve the sizes, distances, or orientations, but all different designs preserve topology even when the map is rotated or mirrored—in all maps the Alps are in Europe and Australia is disjoint from Africa.

The Region Connection Calculus (RCC) by Randell, Cui, and Cohn [138] is probably the best known approach to qualitative spatial representation and reasoning. RCC is a fully axiomatized first-order theory for representing topological relationships between spatial entities.   In doing so, it does not distinguish between different kinds of spatial entities: all spatial entities are regarded as spatial regions. Physical objects in space are represented by the spatial regions they cover. In RCC, spatial regions are defined as regular subsets of some topological space $\mathcal{U}$. Therefore, all spatial regions have the same dimension as $\mathcal{U}$ and are allowed to consist of different disconnected pieces. RCC is based on a single primitive relation, the binary "connected" relation $C(x, y)$ which is true if the topological closures of the spatial regions x and y are connected, i.e., if they have a common point. Other relations can be defined as expressions in first-order logic using C. This enables us to formally define many different relations. Since first-order logic is undecidable, this does not lead to decidable reasoning mechanisms for RCC. In fact, undecidability of the RCC-theory follows from a result by Grzegorczyk [78].

In qualitative spatial representation and reasoning, we are interested in a set of base relations. Randell et al. [138] suggested a set of eight topological base relations definable in the RCC-theory as being of particular importance. This set, which was later called RCC-8, contains as base relations "x is disconnected from y", "x is externally connected to y", "x partially overlaps y", x is equal to y", "x is a tangential proper part of y", "x is a non-tangential proper part of y", and the converse of the latter two relations (x and y are spatial regions). In addition, RCC-8 contains all possible unions of the base relations. The topological distinctions made by this set of eight base relations correspond to those made by Allen's well known interval algebra [4] mainly used for qualitative temporal representation and reasoning.   Thus, RCC-8 can be regarded as the spatial counterpart of the temporal (one-dimensional) interval algebra. The same set of topological relations was independently proposed by Egenhofer [44] as being important for geographical information systems (though Egenhofer defines them using different methods and places stronger constraints on the domain of spatial regions, e.g., regions must be two-dimensional, one-piece, and are not allowed to have holes). RCC-8 is one of the smallest sets of topological base relations which makes topological distinctions rather than just mereological ones. Most other relations definable in the RCC-theory are refinements of these relations. As such RCC-8 is ideally suited as a starting point for qualitative spatial representation and reasoning which can be further extended by other topological distinctions and also by other spatial aspects. As another interesting property of RCC-8, the formal semantics of these relations can be described by using propositional logics rather than first-order logics [10, 12]. Thus, reasoning over the RCC-8 relations is decidable.

Despite the popularity of the Region Connection Calculus and in particular of the RCC-8 relations, little is known about the computational properties of reasoning over these relations. A disappointing result was obtained by Grigni et al. [77] who showed that reasoning with Egenhofer's base relations is NP-hard. This, however, is due to Egenhofer's restrictions on the domain of spatial regions, and is, thus, not transferable to reasoning over RCC-8. The contrary result is by Nebel [126] who found that reasoning over the RCC-8 base relations is tractable. For his proofs, Nebel used Bennett's encoding of RCC-8 in propositional intuitionistic logic [10] which does not enforce regularity of spatial regions. Adding the regularity condition makes the proofs inapplicable, so, the computational complexity of reasoning over RCC-8 is still unknown. Compared to this, there is a large amount of work on the computational properties of the temporal interval algebra (e.g., [161, 111, 155, 69, 110, 128, 112, 115, 157, 127, 41, 42]).

As most of these results were obtained by exploiting either the endpoint representation of temporal intervals or properties of one-dimensional space, neither the results about the interval calculus nor the methods for achieving them can be transferred to the RCC-8 calculus.

1.5 Overview of This Book

In this work we provide a rigorous analysis of the topological RCC-8 calculus for qualitative spatial representation and reasoning. In contrast to previous work on temporal reasoning we try to keep the methods we develop and our proof techniques as general as possible. This enables us to apply some of them also to other systems of relations, be it spatial, temporal, or any other form of relational reasoning. Thus, this book should not only be of interest to people working on RCC-8 or qualitative spatial representation and reasoning, but also to a wider audience interested in constraint satisfaction, efficient algorithms, or cognitive adequacy.

The book is structured as follows. We start with a concise summary of the necessary background information (Chapter 2). This includes sections on topology, on logics, on computational complexity, on constraint satisfaction, and on temporal reasoning with Allen's interval relations. In Chapter 3 we summarize the state of the art in qualitative spatial representation and reasoning followed by a detailed overview of the Region Connection Calculus (Chapter 4) with a focus on RCC-8. After presenting empirical investigations on the cognitive adequacy of RCC-8 (Chapter 5), we analyze the computational properties of reasoning over RCC-8. In Chapter 6 we prove NP-hardness of the consistency problem of RCC-8, identify a large maximal tractable subset of RCC-8, and show that enforcing path-consistency is sufficient for deciding consistency of constraints over relations of this set. In Chapter 7 we propose a general method for proving tractability of reasoning over systems of binary relations. Using this method we make a complete analysis of tractability by identifying two large new maximal tractable subsets of RCC-8 that contain all base relations and for which path-consistency is sufficient for deciding consistency. The maximal tractable subsets can be used to speed up reasoning with RCC-8 which is shown in Chapter 8 in a detailed empirical evaluation.

In Chapter 9 we are concerned with representational properties of RCC-8. We give a canonical model by which all spatial configurations can be represented and show that every consistent set of RCC-8 constraints can be realized in topological spaces of arbitrary dimension. Finally, Chapter 10 summarizes our contributions and gives an outlook to further research.

In the following we give a more detailed overview of this work and of our contributions. We start with an empirical assessment of the RCC-8 relations in order to find out whether they can be considered as cognitively adequate. Although there seems to be a natural agreement of what topological distinctions are important (the RCC-8 relations were independently suggested as significant by Randell et al. [138] and by Egenhofer [44]), there are so far no empirical or cognitive justifications for their importance. In our empirical evaluation we perform a grouping task in which subjects are asked to group items showing different spatial configurations according to their similarity. It turns out that most subjects (1) grouped the items with respect to the RCC-8 relations and (2) made further distinctions only as a refinement of the RCC-8 relations. These results give strong evidence that the distinctions made by the RCC-8 relations can be regarded as cognitively adequate on empirical grounds. Thus, the distinctions made by the RCC-8 relations are natural in the sense that they are also regularly made by humans and, hence, are easy to communicate. Therefore, RCC-8 is an interesting calculus for qualitative spatial representation and reasoning which is worth being studied in detail. The second result of the empirical investigation makes RCC-8 particularly interesting as a starting point for approaching the goal of having an expressive calculus that covers many different aspects of space.

Our main concern in this work are the computational properties of RCC-8. In particular we try to find the most efficient algorithms for several reasoning tasks in order to make qualitative spatial representation and reasoning with RCC-8 as widely applicable as possible. As already mentioned above, checking consistency of a set of constraints is the main reasoning task to which many other reasoning tasks can be polynomially reduced to. By reducing an NP-complete propositional satisfiability problem, we found the consistency problem to be NP-hard in general, i.e., if constraints over all 256 RCC-8 relations can be used. This leads to the question of whether there are subsets of RCC-8 for which the consistency problem can be decided in polynomial time. The goal of this search is to find maximal tractable subsets of RCC-8, i.e., subsets for which deciding consistency is tractable if only constraints over this subset are used, and for which it becomes intractable if constraints over other relations not contained in the set are added. Given such a maximal tractable subset it is not only possible to reason efficiently about constraints over these relations, but also to decide instances of the general NP-hard consistency problem faster by backtracking over sub-instances containing only relations of the maximal tractable subset. This reduces the search space considerably. In order to do so, it is necessary that all base relations are contained in the maximal tractable subset. We are, therefore, only interested in maximal tractable subsets that contain all base relations.

In order to identify a tractable subclass of RCC-8 we use a modification of Bennett's encoding of RCC-8 in modal logic [12] and show that whenever a set of constraints over RCC-8 is consistent, there is a particular Kripke model of the corresponding modal encoding. We use this Kripke model for transforming the modal encoding of RCC-8 into an equivalent propositional formula of polynomial size. As a side effect, this transformation demonstrates that the RCC-8 consistency problem is in NP, but mainly it enables us to identify those relations of RCC-8 that are transformed to propositional Horn formulas—a total number of 148 different relations. Since the satisfiability problem of propositional Horn formulas is tractable, the consistency problem of the set of these 148 relations is also tractable. Using a computer-assisted case analysis, we show that this set, which is denoted by $\widehat{\mathcal{H}}_8$, is a maximal tractable subset of RCC-8. By relating positive unit resolution, a complete method for deciding satisfiability of Horn formulas, to the path-consistency method, we show that enforcing path-consistency is sufficient for deciding consistency of sets of constraints over $\widehat{\mathcal{H}}_8$. Using the same technique, we identify an efficient method for finding a consistent scenario, i.e., a consistent refinement of any consistent set of constraints over $\widehat{\mathcal{H}}_8$ to constraints over the RCC-8 base relations.

Having identified a maximal tractable subset of RCC-8 that contains all base relations, the question is whether there are any other such sets. Instead of trying to find transformations of all possibly tractable subsets of RCC-8 to known tractable problems, we propose a general method for proving tractability of reasoning with sets of relations. This method proves sufficiency of path-consistency for deciding consistency of a set of binary relations $\mathcal{S}$ and, hence, tractability of $\mathcal{S}$ by checking whether every relation of a path-consistent set of constraints over $\mathcal{S}$ can always be consistently refined to a relation of a set $\mathcal{T}$ for which path-consistency is known to decide consistency. In order to identify other maximal tractable subsets of RCC-8 we first generate all possible candidates, i.e., subsets containing all base relations which are not known to be NP-hard but which become NP-hard if any other relation is added to the sets. Surprisingly, there are only three possible candidates (one of them is $\widehat{\mathcal{H}}_8$). By using the new method, we prove all three to be tractable. Thus, we have identified all maximal tractable subsets of RCC-8 that contain all base relations. We further use the new method for identifying the fastest possible algorithms for finding a consistent scenario of any consistent set of constraints over the maximal tractable subsets. As a way of showing generality of the new method, we apply it to Allen's interval algebra and derive the known maximal tractable subset [128].

After having made a complete analysis of tractability in RCC-8, we are now interested in how well the NP-complete consistency problem can be solved and in whether and to what degree the maximal tractable subsets can be used to speed up reasoning. In an empirical evaluation of reasoning with RCC-8, we combine different heuristics with different tractable subsets and compare these strategies by running all of them on the same randomly generated instances. Our first observation is that the randomly generated instances show a phase-transition behavior. Almost all of the difficult instances are located in the phase-transition region, while almost all instances outside the phase-transition region are easily solvable by each of our strategies. It turns out that the strategies using one of the maximal tractable subsets for the backtracking search are much faster than the other strategies and also solve more instances within a maximal number of visited nodes in the search space. We tried to solve those instances which cannot be solved by all strategies by using the method of orthogonally combining the different strategies, i.e., running all strategies in parallel. This method, which has already been used in temporal reasoning with Allen's relations [127], is very successful in solving the hard instances of the phase-transition region of the RCC-8 consistency problem—up to a certain size almost all of these instances can be solved very efficiently. Based on this observation we identified a very effective orthogonal combination of different heuristics (mainly involving the maximal tractable subsets) by which almost all instances up to a very large size (about $n = 400$ regions) can be solved in reasonable time—despite the enormous size of the search space. When randomly generating instances involving only relations which are not contained in the maximal tractable subsets, the instances are on average much more difficult than instances involving all RCC-8 relations. Many of these instances cannot be solved efficiently when they are located in the phase-transition region and are larger than a particular size (about $n = 60$ regions). However, the strategies using one of the maximal tractable subsets, in particular $\widehat{\mathcal{H}}_8$, for the backtracking search are also more successful for these instances than the other strategies.

So far we have analyzed cognitive and computational properties of RCC-8, but there are also some important representational aspects of RCC-8 which have not been studied so far. For instance, it is not known how arbitrary spatial regions as used by RCC-8 can be represented in a computational framework. Furthermore, given a consistent set of constraints over RCC-8, the problem of how to find regions that (1) are related as given by the constraints and (2) have a particular dimension is still unsolved. The second property is particularly important because spatial representation and reasoning is mainly used for two- or three-dimensional space while RCC-8 does not distinguish between different dimensions. Without having solved these problems, it is not clear if RCC-8 can be adequately used for two- or three-dimensional space as well. Using the above mentioned Kripke model, we identify a canonical model of RCC-8 by which all consistent sets of constraints can be modeled. We give a transformation of this model to topological space resulting in a simple topological representation of regions with respect to their mutual relationships. This transformation works for topological spaces of arbitrary dimension, thus, consistent sets of RCC-8 constraints can be modeled in any dimension. If all spatial regions must be internally connected, i.e., consist of only one piece, we show that this is also possible for topological spaces of dimension greater or equal to three, while it is already known that it is not possible in two-dimensional space [77].

2. Background

Artificial Intelligence is a highly diversified discipline with influences from different "classical" areas such as Computer Science, Logic, Engineering, Psychology, and Philosophy. Research in Artificial Intelligence has developed a large library of methods and tools. For reaching certain goals or for illuminating different facets of a problem, however, it is often necessary to apply methods and tools borrowed from other fields.

Qualitative spatial representation and reasoning, a recent sub-discipline of Artificial Intelligence, is an example of such an inherently interdisciplinary field: when dealing with space, it is very convenient and helpful to use the large body of work developed in areas like geometry and topology. One of the motivations for qualitative reasoning and representation is that it is considered to be more "human-like" or "cognitively adequate" than classical approaches. Verifying this claim requires applying methods used in Cognitive Psychology. Apart from these methods, qualitative spatial reasoning also requires some of the classical tools used in knowledge representation and reasoning. Amongst them are different kinds of logics for representing knowledge, computational complexity for analyzing the computational properties of reasoning problems, constraint satisfaction as a particular way of representing and reasoning about knowledge, and empirical methods for analyzing the practical behavior of reasoning problems.

In this chapter we will give brief summaries of the methods, tools, and the terminology of the different fields that are used in this work. We will mainly restrict ourselves to introducing things that are necessary for this work and give pointers to the literature for further references.

2.1 Topology

In the area of qualitative spatial representation it is intended to describe properties of and relationships between spatial entities such as regions or points of a certain space, for instance, of a two- or three-dimensional Euclidean space. Topology offers a rich theory of space by categorizing different kinds of spaces, so-called topological spaces, according to different properties. If one wants to develop a calculus for qualitative spatial reasoning which does

J. Renz: Qualitative Spatial Reasoning with Topological Information, LNAI 2293, pp. 13–29, 2002.
© Springer-Verlag Berlin Heidelberg 2002

not depend on a particular space but can be applied to a more general notion of space, then topology is a powerful tool.

In this section we introduce and define the topological concepts that are used in this work. This includes the notion of a topological space, different kinds of regions such as open, closed, regular open, and regular closed regions, different parts of regions such as the interior, the exterior and the boundary of a region, as well as neighborhoods, neighborhood systems, and different kinds of points. These concepts are very basic and can be found in this or in a similar form in any book on general topology or point-set topology (e.g., [125, 9]). We start with the formal definition of a topology and a topological space:

Definition 2.1 (topology, topological space). *Let $\mathcal{U}$ be a non-empty set, the* universe. *A topology on $\mathcal{U}$ is a family T of subsets of $\mathcal{U}$ that satisfies the following axioms:*

1. *$\mathcal{U}$ and $\emptyset$ belong to T,*
2. *the union of any number of sets in T belongs to T,*
3. *the intersection of any two sets of T belongs to T.*

A topological space is a pair $\langle \mathcal{U}, T \rangle$. The members of T are called open *sets.*

In a topological space $\langle \mathcal{U}, T \rangle$, a subset X of $\mathcal{U}$ is called a *closed* set if its complement X^c is an open set, i.e., if X^c belongs to T. By applying the DeMorgan laws, we obtain the properties of closed sets:

1. $\mathcal{U}$ and $\emptyset$ are closed sets,
2. the intersection of any number of closed sets is a closed set,
3. the union of any two closed sets is a closed set.

If the particular topology T on a set $\mathcal{U}$ is clear or not important, then $\mathcal{U}$ can be referred to as the topological space.

Closely related to the concept of an open set is that of a neighborhood.

Definition 2.2 (neighborhood, neighborhood system). *Let $\mathcal{U}$ be a topological space and $p \in \mathcal{U}$ be a point in $\mathcal{U}$.*

- *$N \subset \mathcal{U}$ is said to be a* neighborhood *of p if there is an open subset $O \subset \mathcal{U}$ such that $p \in O \subset N$.*
- *The family of all neighborhoods of p is called the* neighborhood system *of p, denoted as $\mathcal{N}_p$.*

A neighborhood system $\mathcal{N}_p$ has the property that every finite intersection of members of $\mathcal{N}_p$ belongs to $\mathcal{N}_p$. Based on the notion of neighborhood it is possible to define certain points and areas of a region.

Definition 2.3 (interior, exterior, boundary, closure). *Let $\mathcal{U}$ be a topological space, $X \subset \mathcal{U}$ be a subset of $\mathcal{U}$ and $p \in \mathcal{U}$ be a point in $\mathcal{U}$.*

– p *is said to be an* interior point *of* X *if there is a neighborhood* N *of* p *contained in* X. *The set of all interior points of* X *is called the* interior *of* X, *denoted* $i(X)$.

– p *is said to be an* exterior point *of* X *if there is a neighborhood* N *of* p *that contains no point of* X. *The set of all exterior points of* X *is called the* exterior *of* X, *denoted* $e(X)$.

– p *is said to be a* boundary point *of* X *if every neighborhood* N *of* p *contains at least one point in* X *and one point not in* X. *The set of all boundary points of* X *is called the* boundary *of* X, *denoted* $b(X)$.

– *The* closure *of* X, *denoted* $c(X)$, *is the smallest closed set which contains* X.

The closure of a set is equivalent to the union of its interior and its boundary. Every open set is equivalent to its interior, every closed set is equivalent to its closure.

Definition 2.4 (regular open, regular closed). *Let* X *be a subset of a topological space* $\mathcal{U}$.

– X *is said to be* regular open *if* X *is equivalent to the interior of its closure, i.e.,* $X = i(c(X))$.

– X *is said to be* regular closed *if* X *is equivalent to the closure of its interior, i.e.,* $X = c(i(X))$.

Two sets of a topological space are called *separated* if the closure of one set is disjoint from the other set, and *vice-versa*. A subset of a topological space is *internally connected* if it cannot be written as a union of two separated sets.

Topological spaces can be categorized according to how points or closed sets can be separated by open sets. Different possibilities are given by the *separation axioms* T_i. A topological space $\mathcal{U}$ that satisfies axiom T_i is called a T_i space. Three of these separation axioms which are important for this work are the following:

T_1 : Given any two distinct points $p, q \in \mathcal{U}$, each point belongs to an open set which does not contain the other point.

T_2 : Given any two distinct points $p, q \in \mathcal{U}$, there exist disjoint open sets $O_p, O_q \subseteq \mathcal{U}$ containing p and q respectively.

T_3 : If X is a closed subset of $\mathcal{U}$ and p is a point not in X, there exist disjoint open sets $O_X, O_p \subseteq \mathcal{U}$ containing X and p respectively.

A *connected space* is a topological space which cannot be partitioned into two disjoint open sets, a topological space is *regular*, if it satisfies axioms T_2 and T_3.

2.2 Propositional and First-Order Logics

Logics are a very powerful tool for expressing facts, rules, or possible interconnections of those, e.g., "Paris is closer to Freiburg than London", or "everything which is inside A must be smaller than A", or "if Peter is standing right opposite to Paul, then everything which is to the left of Peter is to the right of Paul". Many different logical formalisms have been developed in Computer Science, Mathematics, and Philosophy, and they are often used in knowledge representation and reasoning in particular and in Artificial Intelligence in general.

A logical formalism consists of three parts, the syntax of logical formulas, the semantics of those formulas, i.e., how logical formulas can be interpreted, and inference rules, i.e., how new formulas can be derived from given formulas. In this work, first-order logics are used to define spatial relations and properties of these relations. Propositional modal logics are used to obtain a simpler encoding of certain sets of spatial relations. This encoding enables us to apply the semantics of the modal formalism to the spatial relations, which is a great advantage over using the semantics of the first-order formalism: the propositional logics we are using are decidable while first-order logics are usually undecidable. The encoding of spatial relations in classical propositional logic is mainly used for deriving complexity results about sets of spatial relations.

In the following we introduce classical propositional logic, propositional modal logics, and first-order logics. Instead of distinguishing between classical propositional logic and propositional modal logics, we refer to classical propositional logic simply as "propositional logic" and to propositional modal logics simply as "modal logics".

2.2.1 Propositional Logic

In *propositional logic* [63, 60] the basic elements are *propositions* also called *atoms* which can have two mutually exclusive *truth values*, *true* and *false*. A *propositional formula* ϕ can be an atom, the *negation* of a propositional formula ϕ_1, written as $\neg\phi_1$, the *disjunction* of two propositional formulas ϕ_1, ϕ_2, written as $\phi_1 \vee \phi_2$, the *conjunction* of two propositional formulas ϕ_1, ϕ_2, written as $\phi_1 \wedge \phi_2$. An *implication* of two propositional formulas, written as $\phi_1 \to \phi_2$, can be expressed equivalently as a combination of negation and disjunction of ϕ_1 and ψ_2. $\psi_1 \to \psi_2 = \neg\phi_1 \vee \phi_2$.

A *truth assignment* T is a mapping of a finite set of atoms to the truth values *true* or *false*. A truth assignment T *satisfies* a propositional formula ϕ, written as $T \models \phi$, if and only if the truth value of ϕ is *true*. The truth value of a propositional formula ϕ is defined inductively on the structure of ϕ. Atoms and negations of atoms are called *literals*. In this respect, atoms are called *positive literals* and negations of atoms are called *negative literals*. We will sometimes distinguish between literals and the single occurrences of literals in a propositional formula, referred to as *literal occurrences*.

Every propositional formula ϕ can be written in an equivalent *normal form* ϕ_N, i.e., $T \models \phi$ if and only if $T \models \phi_N$. The *disjunctive normal form* (DNF) ϕ_{DNF} of a propositional formula ϕ has the structure $\phi_{DNF} = \bigvee_{i=1}^{n} C_i$, where $n \geq 1$ and each C_i is a conjunction of literals. The *conjunctive normal form* (CNF) ϕ_{CNF} of a propositional formula ϕ has the structure $\phi_{CNF} = \bigwedge_{i=1}^{n} C_i$, where $n \geq 1$ and each C_i is a disjunction of literals. A disjunction of literals is called a *clause*. A clause can be regarded as a set of literals and a CNF formula can be regarded as a set of clauses. A clause with only one literal is called a *unit clause*, if the literal is positive, then it is called a *positive unit clause*, if the literal is negative, then it is called a *negative unit clause*. A clause with exactly two literals is called a *Krom clause*, a clause with at most one positive literal is called a *Horn clause*. Formulas consisting of only these clauses are called *Krom formulas* and *Horn formulas*, respectively. A Horn clause without a positive literal is called *indefinite Horn clause*, otherwise it is a *definite Horn clause*.

Given a propositional formula ϕ, an important question is whether there is a truth assignment that satisfies ϕ, i.e., whether ϕ is *satisfiable*. If any truth assignment satisfies ϕ, then ϕ is *valid* or a *tautology*. If no truth assignment satisfies ϕ, then ϕ is *unsatisfiable*. It is clear that a propositional formula ϕ is unsatisfiable if and only if its negation is valid. One method for checking unsatisfiability is *resolution*. Using resolution it is possible to derive new clauses φ from an initial set of clauses ϕ such that if ϕ is satisfiable then $\phi \cup \varphi$ is also satisfiable. The *resolution principle* is the following: Given a set of clauses ϕ and two clauses $\varrho, \varsigma \in \phi$ with $\varrho = (L \vee \varrho')$ and $\varsigma = (\neg L \vee \varsigma')$ where L and $\neg L$ are opposing literals of the same atom, then ϱ and ς resolve to produce the new clause $\varphi = \varrho' \vee \varsigma'$, the *resolvent* of ϱ and ς. If and only if the *empty clause*, i.e., the clause containing no literal, can be derived from a set of clauses ϕ by successively applying the resolution principle, then ϕ is unsatisfiable. Resolution is a sound and complete method for deciding satisfiability of a set of clauses. The worst case running time for resolution is exponential in the number of literals.

A special resolution strategy which is important in this work is *unit resolution*, where always one of the two resolving clauses must be a unit clause. This has the advantage that the resolvent is always smaller than the parent clauses which focuses the search towards producing the empty clause. Unfortunately, unit resolution is not complete for arbitrary sets of clauses. However, unit resolution is complete for sets of Horn clauses. *Positive unit resolution* (PUR) is a variant of unit resolution where always one of the two resolving clauses must be a positive unit clause. This is still complete for Horn clauses.

2.2.2 Propositional Modal Logics

Propositional modal logics [93, 51, 21] have the same syntax as classical propositional logic except for an additional unary operator $\Box$. The modal logics which are important for this work are the so-called *normal modal logics*, i.e., the family of logics that are obtained by extending the basic normal modal logic K. K contains all classical propositional tautologies, the axiom schema

$\Box(\phi \to \psi) \to (\Box\phi \to \Box\psi)$, and is closed under the following inference rules: *modus ponens* (if $\phi, \phi \to \psi \in K$ then $\psi \in K$), *uniform substitution* (if $\phi \in K$ and p occurs in ϕ then $\phi[\psi/p] \in K$ for any ψ), and the *rule of necessitation* (if $\phi \in K$ then $\Box\phi \in K$).

Modal formulas are usually interpreted by means of *Kripke semantics*. A *Kripke model* $\mathcal{M} = \langle W, R, \nu \rangle$ is built upon a *frame* and a *valuation*. A frame $\mathcal{F} = \langle W, R \rangle$ consists of a set of *worlds* W together with an *accessibility relation* $R \subseteq W \times W$. A valuation ν assigns truth values to all the propositional atoms in every world. If a world $v \in W$ is accessible from a world $w \in W$, i.e. $(w, v) \in R$, we say that v is an *R-successor* of w.

The truth value of a modal formula ϕ in a world w of a model $\mathcal{M}$, written as $\mathcal{M}, w \Vdash \phi$, is defined inductively on the structure of ϕ:

$\mathcal{M}, w \Vdash a$ for an atom a	iff	$\nu(w, a) = $ true
$\mathcal{M}, w \Vdash \neg\phi$	iff	$\mathcal{M}, w \not\Vdash \phi$
$\mathcal{M}, w \Vdash \phi \wedge \psi$	iff	$\mathcal{M}, w \Vdash \phi$ and $\mathcal{M}, w \Vdash \psi$
$\mathcal{M}, w \Vdash \phi \vee \psi$	iff	$\mathcal{M}, w \Vdash \phi$ or $\mathcal{M}, w \Vdash \psi$
$\mathcal{M}, w \Vdash \phi \to \psi$	iff	$\mathcal{M}, w \not\Vdash \phi$ or $\mathcal{M}, w \Vdash \psi$
$\mathcal{M}, w \Vdash \Box\phi$	iff	for all u with wRu: $\mathcal{M}, u \Vdash \phi$

Note that the modal operator $\Box$ is related to the accessibility relation R (see [158]).

Other normal modal logics are obtained by extending K with axioms that formalize properties of R. Some well-known examples of modal axioms and corresponding constraints on the accessibility relation are given in the following table:

Name	Axiom	Constraint on R
K	$\Box(\phi \to \psi) \to (\Box\phi \to \Box\psi)$	–
T	$\Box\phi \to \phi$	reflexive
4	$\Box\phi \to \Box\Box\phi$	transitive
5	$\neg\Box\phi \to \Box\neg\Box\phi$	Euclidean
		(wRu and wRv implies uRv)

Two modal logics, which are of particular interest in this work, are S4 and S5. S4 is the extension of the modal logic K by **T** and **4**, closed under modus ponens, uniform substitution, and the rule of necessitation. This is equivalent to specifying that the accessibility relation is reflexive and transitive. S5 is a similar extension of K by **T**, **4**, and **5**, which is equivalent to specifying that the accessibility relation is an equivalence relation. The modal operator is named according to the modal logic, so, e.g., the operator of an S5-frame is called the S5-operator.

Multi-modal logics contain more than one modal operator $\Box$. Each different $\Box_i$ is associated to a different accessibility relation $R_i \subseteq W \times W$, i.e.,

$\mathcal{M}, w \Vdash \Box_i\phi$ iff   for all u with wR_iu: $\mathcal{M}, u \Vdash \phi$.

2.2.3 First-Order Logic

First-order logic [63, 36] is an extension of propositional logics (also known as *0-order logics*) by *variables* x, y, z, *function constants* f, g, h, *relation constants* P, Q, R, and *quantifiers* $\forall, \exists$. These additional elements make first-order logics much more expressive than propositional logics. The syntax of first-order logic is defined inductively. A *term* is either a variable, a proposition, also denoted *object constant*, or a *functional expression* $f(t_1, \ldots, t_n)$ which consists of an n-ary function constant f and n terms $t_1, \ldots, t_n$. An object constant can be regarded as a 0-ary function constant. In first-order logic, an *atom* $P(t_1, \ldots, t_n)$ consists of an n-ary relation constant P and n terms $t_1, \ldots, t_n$. A *first-order formula* ϕ can be an atom, the negation of a first-order formula, the disjunction and conjunction of a first-order formula, or a *quantified* first-order formula which can either be an *existentially quantified* first order formula $\exists x.\phi_1$ or a *universally quantified* first-order formula $\forall x.\phi_1$ where x is a variable and ϕ_1 is a first-order formula.

The semantics of terms and first-order formulas is determined by the interpretation of the symbols used. An *interpretation* $\mathcal{I}$ is a pair $\langle \mathcal{U}, \mathcal{A} \rangle$ formed by a non-empty set $\mathcal{U}$ and a mapping $\mathcal{A}$. $\mathcal{A}$ maps every n-ary function constant f to an n-ary function $\mathcal{A}(f) : \mathcal{U}^n \mapsto \mathcal{U}$ and every n-ary relation constant P to an n-ary relation $\mathcal{A}(P) \subseteq \mathcal{U}^n$. This mapping is continued for terms, i.e., $\mathcal{A}(f(t_1, \ldots, t_n)) = \mathcal{A}(f)(\mathcal{A}(t_1), \ldots, \mathcal{A}(t_n))$. A *variable assignment* α assigns variables with elements of $\mathcal{U}$. $\alpha[x/d]$ means that α assigns the variable x with $d \in \mathcal{U}$.

The truth value of a first-order formula ϕ under an interpretation $\mathcal{I}$ and a variable assignment α is defined inductively on the structure of ϕ. If ϕ is true under $\mathcal{I}$ and α, then $\mathcal{I}$ under α is a *model* of ϕ, written as $\mathcal{I}, \alpha \models \phi$:

$$
\begin{array}{lll}
\mathcal{I}, \alpha \models P(t_1, \ldots, t_n) & \text{iff} & \langle \mathcal{A}(t_1), \ldots, \mathcal{A}(t_n) \rangle \in \mathcal{A}(P) \\
\mathcal{I}, \alpha \models \neg\phi & \text{iff} & \mathcal{I}, \alpha \not\models \phi \\
\mathcal{I}, \alpha \models \phi \wedge \psi & \text{iff} & \mathcal{I}, \alpha \models \phi \text{ and } \mathcal{I}, \alpha \models \psi \\
\mathcal{I}, \alpha \models \phi \vee \psi & \text{iff} & \mathcal{I}, \alpha \models \phi \text{ or } \mathcal{I}, \alpha \models \psi \\
\mathcal{I}, \alpha \models \phi \rightarrow \psi & \text{iff} & \mathcal{I}, \alpha \not\models \phi \text{ or } \mathcal{I}, \alpha \models \psi \\
\mathcal{I}, \alpha \models \forall x.\phi & \text{iff} & \text{for all } d \in \mathcal{U}: \mathcal{I}, \alpha[x/d] \models \phi \\
\mathcal{I}, \alpha \models \exists x.\phi & \text{iff} & \text{there is a } d \in \mathcal{U} \text{ such that } \mathcal{I}, \alpha[x/d] \models \phi
\end{array}
$$

2.3 Computational Complexity

Meanwhile computers have entered almost every part of our life and we expect them to work accurately and efficiently—real-time. Consider, for instance, a car navigation system which computes the optimal path for reaching a goal as fast as possible and which gives an alternative path if some unexpected situation occurs. This system must give the right direction before every crossing. It cannot be accepted to stop and wait until the computation is finished or to receive the correct direction after a crossing. If solving a computational problem is too inefficient, two possible solutions are to develop more efficient algorithms or faster hardware (or accept a fast approximate solution). However, because of their inherent complexity, many computational problems cannot be solved efficiently and even substantially faster hardware does not help. Therefore, it is important to know the complexity of a problem for designing the best possible algorithms.

In the field of computational complexity [130], computational problems are classified according to their need for ressources for solving them, usually the running time and the memory consumption. This allows to compare the complexity of different problems and to design algorithms for a whole class of problems. For classifying computational problems, they are usually expressed as *decision problems*, i.e., problems that require a simple yes/no answer. Most problems can be easily translated into an equivalent decision problem. An example for a decision problem is the propositional satisfiability problem SAT: given a set of variables V and a propositional formula ϕ over V in CNF, is there a satisfying truth assignment for ϕ? The complexity of a decision problem is usually measured according to the worst-case running time or memory consumption of the best possible algorithm.

Running time as well as memory consumption of an algorithm depends on the size n of its input, i.e., on the size of the problem instance, and can be expressed as a function $f(n)$. For classifying algorithms according to their running time, the asymptotical behavior is more important than f itself. This is specified in terms of the *O-notation* which gives an upper bound on the running time within a constant factor [32]. An algorithm with a running time of $O(n^3)$ or faster is usually considered to be efficient. In areas like database systems where instances have a very large size, a running time of $O(n^3)$ is too slow. In these areas efficient algorithms should have a linear or at most quadratic running time.

2.3.1 Tractability and NP-Completeness

There is a large number of different complexity classes that are used to categorize decision problems [100]. Particularly important is the class of decision problems that can be solved in polynomial time using a deterministic algorithm. This complexity class is called P. The class of problems solvable in polynomial time using a non-deterministic algorithm is called NP, which is equivalent to specifying that a given solution of an NP problem can be verified in polynomial time using a deterministic algorithm. Unless otherwise stated, an algorithm will always be a deterministic algorithm. It is clear that P is a subset of NP, but it is not known whether P is a proper subset of NP or whether P is equal to NP, which is called the $P \stackrel{?}{=} NP$ problem. Problems in P are also called *tractable* problems, problems outside P are called *intractable* problems.

An important method of comparing problems is specifying a *reduction* from one problem to another. A problem A can be *reduced* to a problem B

by giving a constructive transformation $f : A \mapsto B$ of every instance of A to an equivalent instance of B, i.e., every instance a of A has a solution if and only if $f(a)$ has a solution. If f can be computed in polynomial time, the reduction is a *polynomial (time) reduction*. If A is polynomially reducible to B (written as $A \leq_p B$), then any polynomial time algorithm for solving B can be used to solve A. Thus, for showing that a particular decision problem A is in P, it is sufficient to find another problem $B \in$ P such that $A \leq_p B$.

A decision problem A is said to be NP-*hard* if any other problem in NP can be polynomially reduced to A. An NP-hard problem which is itself contained in NP is called NP-*complete*. NP-complete problems are the most difficult problems in NP. In order to prove a decision problem A to be NP-hard, it is sufficient to find another NP-hard problem that can be polynomially reduced to A. The first problem that was identified to be NP-complete is the SAT problem [31]. In this work we use the following NP-complete propositional decision problems [62]:

Given: A set of variables V and a propositional formula ϕ over V in CNF such that each clause of ϕ has exactly three literals.

Questions:
1. Is there a satisfying truth assignment for ϕ? (3SAT)
2. Is there a satisfying truth assignment for ϕ such that each clause has at least one true literal and at least one false literal? (NOT-ALL-EQUAL-3SAT)
3. Is there a satisfying truth assignment for ϕ such that each clause has exactly one true literal? (ONE-IN-THREE-3SAT)

Some variants of the propositional satisfiability problem are solvable in polynomial time. This includes the 2SAT problem, the propositional satisfiability problem of Krom formulas, and the HORNSAT problem, the propositional satisfiability problem of Horn formulas, which is of particular importance in this work. It is generally believed that P $\neq$ NP, and, hence, that NP-complete problems are intractable. This is also the assumption of this work. So far, any algorithm for an NP-complete problem has an exponential or at least super-polynomial running time.

2.3.2 Phase Transitions

Having proved a problem to be NP-complete is not the end of the computational analysis of a problem, rather its beginning. NP-completeness is just a worst-case measure of a problem. It means that for any algorithm there is at least one instance of an NP-complete problem which cannot be solved in polynomial time. It is possible that only one in a million instances is very hard and that the other instances can be solved efficiently.

There are several ways to deal with NP-complete problems. One way is to develop efficient approximation algorithms which are correct but not complete for deciding either solubility or insolubility. Another way is to use

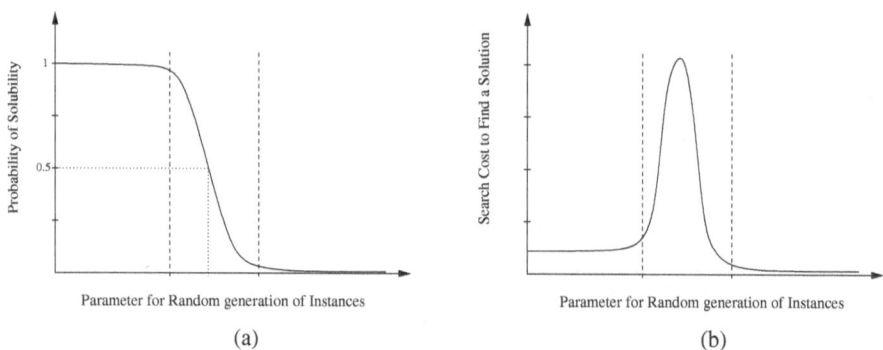

Fig. 2.1. Typical phase-transition behaviour of randomly generated instances

complete algorithms which require exponential time in the worst-case and
to develop heuristics which solve many instances efficiently. In all cases the
effectiveness of new algorithms and heuristics should be verified using a large
number of instances. Since it is usually not easy to obtain a large number
of real-world instances, many researchers generate instances randomly with
respect to different control parameters.

Cheeseman, Kanefsky, and Taylor [20] found that randomly generated
instances of the NP-complete problems they studied had a very special behavior: When ordering these instances according to a particular problem-
dependent parameter, there are three different regions with respect to the
solubility of the instances that occur when changing the parameter. In one
region instances are soluble with a very high probability, in one region instances are unsoluble with a very high probability, and in between these two
regions there is a very small region where the probability of solubility of these
instances changes abruptly from very high to very low (see Figure 2.1(a)).
Cheeseman et al. [20] called this region *phase-transition*. In this region a small
change of the local parameter leads to a large change in the solubility of the
instances.

Cheeseman et al. [20] further found that almost all hard instances are located in the phase-transition region. In general, instances in the phase tran-
sistion are harder than instances in the other two regions (see Figure 2.1(b)).
This is because instances in soluble region are underconstrained while instances in the unsoluble region are overconstrained. In some studies, however, it turned out that some underconstrained instances are particularly hard
[66].

The behavior of randomly generated instances of NP-complete problems
described by Cheeseman et al. was found by many researchers for many NP-
complete problems, although satisfiability problems were the most studied
problems. A typical parameter for satisfiability problems that causes a phase
transition is the ratio of clauses-to-variables. An interesting selection of papers on the topic can be found in [91].

2.4 Constraint Satisfaction

Knowledge about entities or about the relationships between entities is often given in the form of *constraints*. For instance, when trying to place furniture in a room there are certain constraints on the position of the objects. Unary constraints such as "The room is 5 metres in length and 6 in breadth" restrict the domain of single variables, the length and the breadth of the room. Binary constraints like "The desk should be placed in front of the window", ternary constraints like "The table should be placed between the sofa and the armchair", or in general n-ary constraints restrict the domain of 2, 3, or n variables. Problems like these are formalized as a *constraint satisfaction problem* (CSP): Given a set of variables $\mathcal{V}$ over a domain $\mathcal{D}$ and a set Θ of (1- to n-ary) constraints on the variables of $\mathcal{V}$. In this work we restrict ourselves to binary CSPs, i.e., CSPs where only binary constraints are used. Most of the following terms and definitions can be used for arbitrary constraint problems. Binary CSPs can be represented as a *constraint network* which is a labeled digraph where each node is labeled by a variable and each directed edge is labeled by a binary constraint.

A CSP is *consistent* if it has a *solution* which is an assignment of values of $\mathcal{D}$ to the variables of $\mathcal{V}$ in a way that all constraints are satisfied. If the domain of the variables is finite, CSPs can be solved by *backtracking* over the ordered domains of the single variables. Backtracking works by successively instantiating variables with values of the ordered domain until either all variables are instantiated and a solution is found or an inconsistency is detected in which case the current variable is instantiated with the next value of its domain. If all possible instantiations of the current variable lead to an inconsistency, the previous variable becomes the current variable and the process is repeated. Backtracking is in general exponential in the number of variables. Constraints can be processed by propagating the domain restrictions from one variable to the others. If the domain of the variables is infinite, backtracking over the domain or restricting the domain is not very helpful and other methods have to be applied.

2.4.1 Binary Constraint Satisfaction Problems and Relation Algebras

One way of dealing with infinite domains is using constraints over a finite set of binary relations. Ladkin and Maddux [110] formulated binary CSPs as *relation algebras* developed by Tarski [154]. This allows treating binary CSPs with finite and infinite domains in a uniform way. A relation algebra consists of a set of binary relations $\mathcal{R}$ which is closed under several operations on relations and contains some particular relations. The operations are *union* ($\cup$), *intersection* ($\cap$), *composition* ($\circ$), *complement* ($\bar{\cdot}$), and *conversion* ($\cdot^{\smile}$). The formal definition of these operations is the following (x, y are variables, R, S are relations):

$$
\begin{aligned}
\forall x, y : & \quad x(R \cup S)y & \leftrightarrow & \quad xRy \vee xSy, \\
\forall x, y : & \quad x(R \cap S)y & \leftrightarrow & \quad xRy \wedge xSy, \\
\forall x, y : & \quad x(R \circ S)y & \leftrightarrow & \quad \exists z : (xRz \wedge zSy), \\
\forall x, y : & \quad xR^{\smile}y & \leftrightarrow & \quad yRx, \\
\forall x, y : & \quad x\overline{R}y & \leftrightarrow & \quad \neg xRy.
\end{aligned}
$$

Since the union of relations is equivalent to the disjunction of constraints, one often refers to the union of relations as a *disjunction* of relations. In the following we will write sets of relations to denote disjunctions of relations, e.g., $\{R, S\}$ is the disjunction of R and S. The required binary relations are the *empty relation* $\emptyset$ which does not hold between any two members of the domain, the *universal relation* $*$ which holds between any two members of the domain, and the *identity relation* Id which holds between every member of the domain with itself.

We assume that a set of constraints Θ contains one constraint for each pair of variables involved in Θ, i.e., if no information is given about the relation holding between two variables x and y, then the universal constraint $x\{*\}y$ is implicitly contained in Θ. Another assumption that we make is that whenever a constraint xRy is in Θ, its converse $yR^{\smile}x$ is also present.

Determining consistency for CSPs with infinite domains is in general undecidable [89]. A partial method for determining inconsistency of a CSP is the *path-consistency method* which enforces path-consistency on a CSP [123, 118]. A CSP is *path-consistent* if and only if for any consistent instantiation of any two variables, there exists an instantiation of any third variable such that the three values taken together are consistent. It is necessary but not sufficient for the consistency of a CSP that path-consistency can be enforced, i.e., a CSP where path-consistency cannot be enforced is not consistent, but a CSP is not necessarily consistent when path-consistency can be enforced. A naive way to enforce path-consistency on a CSP is to strengthen relations by successively applying the following operation until a fixed point is reached:

$$
\forall k : R_{ij} \leftarrow R_{ij} \cap (R_{ik} \circ R_{kj})
$$

where i, j, k are nodes and R_{ij} is the relation between i and j. The resulting CSP is equivalent to the original CSP, i.e., it has the same set of solutions. If the empty relation occurs while performing this operation the CSP is inconsistent, otherwise the resulting CSP is path-consistent. Provided that the composition of relations can be computed in constant time, more advanced algorithms enforce path-consistency in time $O(n^3)$ where n is the total number of nodes in the graph [119]. Figure 2.2 shows the $O(n^3)$ time path-consistency algorithm by van Beek [155].

2.4.2 Relation Algebras Based on JEPD Relations

Of particular interest are relation algebras that are based on finite sets of *jointly exhaustive and pairwise disjoint* (JEPD) relations. JEPD relations

---

Algorithm: PATH-CONSISTENCY
Input: A set Θ of binary constraints over the variables $x_1, x_2, \ldots, x_n$ of Θ.
Output: **fail**, if Θ is inconsistent; path-consistent set equivalent to Θ,
        otherwise.

1. $Q \leftarrow \{(i, j, k), (k, i, j) \mid i < j, k \neq i, k \neq j\}$;
        (i indicates the i-th variable of Θ. Analogously for j and k)
2. *while* $Q \neq \emptyset$ *do*
3. select and delete a path (i, k, j) from Q;
4.   *if* REVISE(i, k, j) *then*
5.   *if* $R_{ij} = \emptyset$ *then* return **fail**
6.   *else* $Q \leftarrow Q \cup \{(i, j, k), (k, i, j) \mid k \neq i, k \neq j\}$;

Function: REVISE(i, k, j)
Input: three variables i, k and j
Output: true, if R_{ij} is revised; false otherwise.
Side effects: R_{ij} and R_{ji} revised using the operations $\cap$ and $\circ$
               over the constraints involving i, k, and j.

1. oldR := R_{ij};
2. $R_{ij} := R_{ij} \cap (R_{ik} \circ R_{kj})$;
3. *if* (oldR $= R_{ij}$) *then* return *false*;
4. $R_{ji} := R_{ij}^{\smile}$;
5. return *true*.

Fig. 2.2. Van Beek's PATH-CONSISTENCY algorithm

---

are sometimes called *atomic*, *basic*, or *base relations*. We refer to them as base relations. Since any two entities are related by exactly one of the base relations, they can be used to represent definite knowledge with respect to the given level of granularity. Indefinite knowledge can be specified by disjunctions of possible base relations. Given a set of base relations $\mathcal{A}$, the powerset $2^{\mathcal{A}}$ of $\mathcal{A}$, i.e., the set of all possible disjunctions of relations of $\mathcal{A}$, is a relation algebra if it is closed under composition.

For these relation algebras, the universal relation is the disjunction of all base relations. Converse, complement, intersection and disjunction of relations can easily be obtained by performing the corresponding set theoretic operations. Composition of base relations has to be computed using the semantics of the relations. Composition of disjunctions of base relations can be obtained by computing the disjunction of the composition of the base relations. Usually, compositions of the base relations are pre-computed and stored in a *composition table*.

We say that a relation R is a *refinement* of a relation S if and only if $R \subseteq S$. Given, for instance, a disjunction of relations $\{R_1, R_2, R_3\}$, then the relation $\{R_1, R_2\}$ is a refinement of the former relation. This definition carries over to constraints and to sets of constraints. Then, a set of constraints Θ' is a refinement of Θ if and only if the same variables are involved in both the sets, and for every pair of variables x, y, if $xR'y \in \Theta'$ and $xRy \in \Theta$ then

---

Algorithm: CONSISTENCY

Input: A set Θ of binary constraints over the variables $x_1, x_2, \ldots, x_n$ and a subset $\mathcal{S} \subseteq 2^{\mathcal{A}}$ that splits $2^{\mathcal{A}}$ exhaustively and for which CSPSAT($\mathcal{S}$) is decidable.

Output: **true**, iff Θ is consistent.

1. PATH-CONSISTENCY(Θ)
2. *if* Θ contains the empty relation *then* return **false**
3. *else* choose an unprocessed constraint $x_i R x_j$ and
        split R into $S_1, \ldots, S_k \in \mathcal{S}$ such that $S_1 \cup \ldots \cup S_k = R$
4. *if* no constraint can be split *then* return DECIDE(Θ)
5. *for* all refinements S_l $(1 \leq l \leq k)$ *do*
6.   replace xRy with xS_ly in Θ
7.   *if* CONSISTENCY(Θ) *then* return **true**

Fig. 2.3. Backtracking algorithm for deciding consistency

$R' \subseteq R$. Θ' is said to be a *consistent refinement* of Θ if and only if Θ' is a refinement of Θ and both Θ and Θ' are consistent. A *consistent scenario* Θ_s of a set of constraints Θ is a consistent refinement of Θ where all the constraints of Θ_s are singletons, i.e., assertions of base relations.

This work deals with determining consistency of binary constraint satisfaction problems that are based on JEPD relations. Let $\mathcal{A}$ be a finite set of JEPD binary relations and $2^{\mathcal{A}}$ be a relation algebra. The *consistency problem* CSPSAT($\mathcal{S}$) for sets $\mathcal{S} \subseteq 2^{\mathcal{A}}$ over a (possibly infinite) domain $\mathcal{D}$ is defined as follows:

Instance: A set V of variables over a domain $\mathcal{D}$ and a finite set Θ of binary constraints xRy, where $R \in \mathcal{S}$ and $x, y \in V$.

Question: Is there an instantiation of all variables in Θ with values from $\mathcal{D}$ such that all constraints are satisfied?

In the general case CSPSAT is undecidable [89], but in many interesting cases it is possible to prove decidability or even tractability of CSPSAT($\mathcal{S}$).

If CSPSAT is decidable for a certain subset $\mathcal{S} \subseteq 2^{\mathcal{A}}$ then it is possible to decide CSPSAT for other subsets of $2^{\mathcal{A}}$ by using backtracking over the relations of $\mathcal{S}$. This is done by successively selecting refinements of the constraints such that the refinements are contained in $\mathcal{S}$. For example, suppose $\mathcal{S} \subseteq 2^{\mathcal{A}}$ contains the relations $S_1, \ldots, S_n$, the relation $R \in 2^{\mathcal{A}}$ is not contained in $\mathcal{S}$, and $R = \{S_1, S_3, S_4\}$. Then, the constraint xRy can be processed by backtracking over the sub-relations of R which are contained in $\mathcal{S}$, i.e., over $S_1, S_3,$ and S_4. In general, a subset $\mathcal{S} \subseteq 2^{\mathcal{A}}$ *splits* another subset $\mathcal{T} \subseteq 2^{\mathcal{A}}$ *exhaustively* if for every relation T of $\mathcal{T}$ there are refinements $S_1, \ldots, S_k \in \mathcal{S}$ such that $T = \{S_1, \ldots, S_k\}$. The backtracking algorithm given in Figure 2.3 is a generalization of the one proposed by Ladkin and Reinefeld [111] and relies on a set $\mathcal{S}$ that splits $2^{\mathcal{A}}$ exhaustively. Note that a set $\mathcal{S}$ splits $2^{\mathcal{A}}$ exhaustively if and only if $\mathcal{S}$ contains all base relations. $\mathcal{S}$ is called the *split set*. The backtracking algorithm uses a function DECIDE which is a sound and complete decision procedure for CSPSAT($\mathcal{S}$). The (optional) procedure

PATH-CONSISTENCY in line 1 is used as forward-checking and restricts the remaining search space. Nebel [127] showed that this restriction preserves soundness and completeness of the algorithm. If the decision procedure DE-CIDE runs in polynomial time, CONSISTENCY is exponential in the number of constraints of Θ. If enforcing path-consistency is sufficient for deciding CSPSAT($\mathcal{S}$), DECIDE(Θ) in line 4 is not necessary. Instead it is possible to always return true there.

The efficiency of the backtracking algorithm depends on several factors. One of them is, of course, the size of the search space which has to be explored. A common way of measuring the size of the search space is the average *branching factor*, i.e., the average number of branches each node in the search space has. For the backtracking algorithm described in Figure 2.3 this depends on the average number of relations of the split set $\mathcal{S}$ into which a relation has to be split. The less splits in average the better, i.e., it is to be expected that the efficiency of the backtracking algorithm depends on the split set $\mathcal{S}$ and its branching factor. Another factor is how the search space is explored. The backtracking algorithm of Figure 2.3 offers two possibilities of applying heuristics. One is in line 3 where the next unprocessed constraint can be chosen, the other is in line 5 where the next refinement can be chosen. These two choices influence the search space and the path through the search space. Good choices should increase efficiency of the backtracking algorithm.

Other fundamental reasoning problems are the *minimal label problem* CSPMIN, the problem of finding the strongest entailed relation for each pair of variables from a given set of constraints, and the *entailment problem* CS-PENT, i.e., decide whether a particular constraint is entailed by a set of constraints. As it was shown for the corresponding temporal problems (see the next section), the entailment problem, the minimal label problem, and the consistency problem are equivalent under polynomial Turing reductions [161, 69].

2.5 Temporal Reasoning with Allen's Interval Algebra

Temporal reasoning has attracted a lot of attention in the past 15 years and many of the results on constraint satisfaction described in the previous section were obtained in this field. A very simple way of reasoning about time is using time points and representing the possible relationships between them. The *point algebra* is formed from the three different base relations $<$, $>$, and $=$ and all possible disjunctions thereof and describes whether a time point is before, equal, or after another time point [160]. Ladkin and Maddux [109] showed that the path-consistency algorithm is complete for deciding inconsistency of sets of constraints over the point algebra.

Of particular importance is the system of relations introduced by Allen [4] (actually by Nicod in the 1920s [99]) which describes the possible relationships between two convex intervals. Every interval can be represented

Table 2.1. The thirteen base relations of Allen's interval algebra

Interval Base Relation	Symbol	Pictorial Example	Endpoint Relations	
x before y y after x	$\prec$ $\succ$	xxx yyy	$X^- < Y^-,$ $X^+ < Y^-,$	$X^- < Y^+,$ $X^+ < Y^+$
x meets y y met-by x	m m⌣	xxxx yyyy	$X^- < Y^-,$ $X^+ = Y^-,$	$X^- < Y^+,$ $X^+ < Y^+$
x overlaps y y overlapped-by x	o o⌣	xxxx yyyy	$X^- < Y^-,$ $X^+ > Y^-,$	$X^- < Y^+,$ $X^+ < Y^+$
x during y y includes x	d d⌣	xxx yyyyyyyy	$X^- > Y^-,$ $X^+ > Y^-,$	$X^- < Y^+,$ $X^+ < Y^+$
x starts y y started-by x	s s⌣	xxx yyyyyyyy	$X^- = Y^-,$ $X^+ > Y^-,$	$X^- < Y^+,$ $X^+ < Y^+$
x finishes y y finished-by x	f f⌣	xxx yyyyyyyy	$X^- > Y^-,$ $X^+ > Y^-,$	$X^- < Y^+,$ $X^+ = Y^+$
x equals y	$\equiv$	xxxx yyyy	$X^- = Y^-,$ $X^+ > Y^-,$	$X^- < Y^+,$ $X^+ = Y^+$

using the two end points of the interval. When comparing the end points of two intervals according to the relations of the point algebra, two intervals can be related by the 13 different JEPD base relations *before* ($\prec$), *meets* (m), *overlaps* (o), *starts* (s), *during* (d), *finishes* (f), their converse relations $\succ$, m⌣, o⌣, s⌣, d⌣, f⌣, and the identity relation *equal* ($\equiv$). Table 2.1 shows examples of these relations and how they are formed by the corresponding endpoint relations where X^- is the starting point and X^+ the endpoint of interval x. The endpoint relations $X^- < X^+$ and $Y^- < Y^+$ are valid for all relations and have been omitted in the table. The full algebra, known as the *interval algebra*, consists of the 2^{13} possible disjunctions of the base relations and has an NP-complete consistency problem [160].

Vilain and Kautz identified a subclass of the interval algebra which can be expressed in terms of the point algebra. This class, denoted as the "pointisable" subclass $\mathcal{P}$, contains 188 different relations, i.e., about 2% of the full algebra, and has a tractable consistency problem [109]. It was further shown that the path-consistency algorithm is sufficient for deciding inconsistency of constraints over pointisable relations [156].

Nebel and Bürckert [128] identified a larger tractable subclass by writing interval constraints as point constraints in a Horn-like form, i.e., conjunctions of disjunctions of point constraints such that at most one of the disjuncts is a non-negative constraint ($p_i \neq p_j$ is a negative point constraint, all others are non-negative). Nebel and Bürckert called this subclass the *ORD-Horn* subclass $\mathcal{H}$. It contains 868 different interval relations including all base relations, i.e., about 10% of the full algebra. Path-consistency is sufficient for deciding inconsistency of reasoning over $\mathcal{H}$ relations. In computer-assisted proofs, Nebel and Bürckert showed that $\mathcal{H}$ is a *maximal tractable subclass*, i.e., if any other relation is added to $\mathcal{H}$, deciding consistency of constraints over the new set becomes NP-complete. They further showed that $\mathcal{H}$ is the only such set that contains all base relations and can, hence, be used to decide consistency of the full algebra by using the backtracking algorithm described in the previous section.

Drakengren and Jonsson [41, 42] identified further maximal tractable subclasses of the interval algebra that do not contain all base relations. The full classification of all subclasses into those having a tractable or NP-complete consistency problem has been done recently by Krokhin et al [107]. Except for $\mathcal{H}$, however, none of the other maximal tractable subclasses can be used for backtracking over the full algebra. They are, thus, mainly interesting for theoretical reasons rather than for increasing efficiency of complete reasoning algorithms.

Nebel [127] empirically studied reasoning with the interval algebra by randomly generating test instances and solving them using different heuristics [157] and different split sets. Although $\mathcal{H}$ is much larger than $\mathcal{P}$, those heuristics relying on $\mathcal{H}$ as the split set were only slightly more efficient than those relying on the split set $\mathcal{P}$. However, when combining the different heuristics orthogonally, i.e., when running them in parallel, more instances were solved within the same CPU time than for any single heuristic alone. The same observation was made when excluding the heuristics relying on the split set $\mathcal{H}$, i.e., using the "$\mathcal{H}$-heuristics" does not lead to solving many instances that were otherwise insoluble within reasonable time. The main contribution of the $\mathcal{H}$-heuristics in the orthogonal combination was that they were often successful in producing the fastest response.

3. Qualitative Spatial Representation and Reasoning

In this chapter we give an overview of the field of qualitative spatial representation and reasoning. Since a lot of research on that topic has been done, this overview cannot be complete. Instead, we try to give an insight into the lines of research which originated and inspired our work. Some more complete overview articles are by Cohn [27, 26] and Vieu [159]. Although qualitative spatial representation and qualitative spatial reasoning can be regarded as two separate sub-fields (where the former deals with the different forms of spatial knowledge and how this can be formalized and represented in a computational framework, while the latter deals with methods and techniques for reasoning about spatial knowledge and developing efficient algorithms) we will in the following subsume both sub-fields by the term qualitative spatial reasoning or QSR for short. On the one hand this is for reasons of brevity, on the other hand because reasoning about spatial knowledge is certainly not possible without representing it.

We start this chapter with a historical sketch of qualitative spatial reasoning where we survey some research that led to the development of the field. After this we mention some principles of qualitative spatial reasoning which completes what has already been said in the introduction. Finally, we summarize some important approaches to qualitative spatial reasoning with a particular emphasis on formal results.

3.1 History of Qualitative Spatial Reasoning

The ability to deal with commonsense knowledge about the world [34, 90] is important for any intelligent system that acts in the real world and it has been early recognized as one of the central topics of Artificial Intelligence to represent and reason about commonsense knowledge [121]. Space has always been considered to be an important part of commonsense reasoning since the physical world has a spatial extend and all objects which are dealt with are located in space relative to other objects. Early approaches involving commonsense knowledge about the physical world were trying to solve text book physics and math problems, e.g. [19, 17], but it soon turned out that mathematical equations were not sufficient for solving most problems. De Kleer [37] suggested a system which involves both qualitative and quantitative know-

J. Renz: Qualitative Spatial Reasoning with Topological Information, LNAI 2293, pp. 31–40, 2002.
© Springer-Verlag Berlin Heidelberg 2002

ledge and as such marked a starting point for *qualitative physics* [54, 165]. A further milestone towards establishing qualitative physics (or *qualitative reasoning* as it is now called) as an important sub-discipline of Artificial Intelligence was the Naive Physics Manifesto [82, 83, 84] which, among other things, proposes to represent space-time with four-dimensional "histories".

Based on Hayes's histories, Forbus [52] presented a system which reasoned about motion through free space by using again both qualitative and quantitative information.

Later, research effort focussed on purely qualitative representations. This was successful in *qualitative dynamics*, the sub-field of qualitative physics which describes forces that cause systems to change over time. This success was largely due to the possibility of qualitatively describing *quantity spaces* [53], for instance, by exploiting the underlying partial or total order using transitivity or qualitative arithmetic. Conversely, it was conjectured that this is not possible in *qualitative kinematics*, the sub-field of qualitative physics which is concerned with spatial reasoning required by commonsense physics. Forbus, Nielsen, and Faltings [55] claim that "there is no purely qualitative, general-purpose kinematics" (the *poverty conjecture*). Their main argument is "No total order: Quantity spaces don't work in more than one dimension, leaving little hope of concluding much by combining weak information about spatial properties." Meanwhile this pessimistic view is outdated or at least weakened as there has been a lot of work on qualitative spatial reasoning which shows that qualitative information about space can be combined in a useful way by using composition, for instance. Some examples of such work are given in Section 3.3.

3.2 Principles of Qualitative Spatial Reasoning

Qualitative reasoning is an approach for dealing with commonsense knowledge without using numerical computation. Instead, one tries to represent knowledge using a limited vocabulary such as qualitative relationships between entities or qualitative categories of numerical values, for instance, using $\{+, -, 0\}$ for representing real values. An important motivation for using a qualitative approach is that it is considered to be closer to how humans represent and reason about commonsense knowledge. Another motivation is that it is possible to deal with incomplete knowledge. Qualitative reasoning, however, is different from fuzzy computation. While fuzzy categories are approximations to real values, qualitative categories make only as many distinctions as necessary—the granularity depends on the corresponding application.

Two very important concepts of commonsense knowledge are time and space. Time, being a scalar entity, is very well suited for a qualitative approach and, thus, qualitative temporal reasoning has early emerged as a lively sub-field of qualitative reasoning which has generated a lot of research effort and important results (cf. Section 2.5). Space, in turn, is much more complex than time. This is mainly due to its inherent multi-dimensionality which leads to a higher degree of freedom and an increased possibility of describing entities and relationships between entities. This becomes clear when enumerating natural language expressions involving space or time. While temporal expressions mainly describe order and duration (like "before", "during", "long", or "a while") or a personal or general temporal category (like "late" or "morning"), spatial expressions are manifold. They are used for describing, for instance, direction ("left", "above"), distance ("far","near"), size ("large", "tiny"), shape ("oval", "convex"), or topology ("touch", "inside").

It is obvious that most spatial expressions in natural language are purely qualitative.

Although there are doubts that space (because of its multi-dimensionality) can be adequately dealt with by using only qualitative methods (the poverty conjecture [55]), qualitative spatial reasoning has become an active research area. Because of the richness of space and its multiple aspects, however, most work in qualitative spatial reasoning has focussed on single aspects of space. The most important spatial aspects are topology, orientation, and distance. As shown in psychological studies [131], this is also the order in which children acquire spatial notions. Other spatial aspects include size, shape, morphology, and spatial change (motion). Orthogonal to this view is the ontological distinction of which kinds of spatial entities are used. One line of research considers points as the basic entities, another line considers extended spatial entities such as spatial regions as basic entities. While it is easier to deal with points rather than with regions in a computational framework, taking regions as the basic entities is certainly more adequate for commonsense reasoning—eventually, all physical objects are extended spatial entities. Furthermore, if points are required, they can be constructed from regions [16]. A further ontological distinction is the nature of the embedding space. The most common notion of space is n-dimensional continuous space (R^n). But there are also approaches which consider, e.g., discrete [61] or finite space [74].

The most popular reasoning methods used in qualitative spatial reasoning are constraint based techniques adopted from previous work in temporal reasoning (see Section 2.4 for a comprehensive introduction to these techniques). In order to apply these methods, it is necessary to have a set of qualitative binary *base relations* which have the property of being jointly exhaustive and pairwise disjoint, i.e, between any two spatial entities exactly one of the base relations holds. The set of all possible relations is then the powerset of the base relations, i.e., all possible unions of the base relations. Reasoning can be done by exploiting *composition* of relations. For instance, if the binary relation R_1 holds between entities A and B and the binary relation R_2 holds between B and C, then the composition of R_1 and R_2 restricts the possible relationship between A and C. Compositions of relations are usually pre-computed and stored in a *composition table*.

In the following section we summarize some important approaches of qualitative spatial reasoning for dealing with different aspects of space.

3.3 Different Approaches to Qualitative Spatial Reasoning

In qualitative spatial reasoning it is common to consider a particular spatial aspect such as topology, direction, or distance and to develop a system of qualitative relationships between spatial entities which cover this spatial aspect to some degree and which appear to be useful from an applicational or from a cognitive perspective. If these relations are based on a set of jointly exhaustive and pairwise disjoint base relations which is closed under several operations, it is possible to apply constraint based methods for reasoning over these relations (see Section 2.4). For this it is only necessary to give a composition table; either for all relations or for the base relations plus a procedure for computing the compositions of complex relations.

The composition table should be obtained using the formal semantics of the relations. Otherwise it is not possible to verify correctness and completeness of the inferences. Formal semantics of the relations are also necessary for finding efficient reasoning algorithms which are essential for most applications. Without formal semantics it is sometimes not even possible to show that reasoning over a system of relations is decidable (e.g., the Egenhofer relations [44] as shown by Grigni et al. [77]).

In the following subsections we survey some important approaches to the main aspects of space, topology, direction, and distance. Instead of summarizing many different approaches, we focus on those approaches which have been formally analyzed.

3.3.1 Topology

Topological distinctions between spatial entities are a fundamental aspect of spatial knowledge. Topological distinctions are inherently qualitative which makes them particularly interesting for qualitative spatial reasoning. Although there is a large body of work on topology developed in mathematics (see Section 2.1), this is not very well suited for qualitative spatial reasoning. Mathematical research on topology is not concerned with reasoning over topological relationships and as such does not provide us with any reasonable topological calculi and reasoning mechanisms [75].

Topological approaches to qualitative spatial reasoning usually describe relationships between spatial regions rather than points, where spatial regions are subsets of some topological space. Most existing approaches on formalizing topological properties of spatial regions are based on work from Whitehead [166] and Clarke [22, 23] who axiomatized mereotopologies using a single primitive relation, the binary connectedness relation. Some approaches also distinguish between a mereological primitive, the parthood relation, and a topological primitive, the connected relation [18]. Using these primitive relations it is possible to define many other relations. A set of jointly exhaustive and pairwise disjoint relations which can be defined in all approaches of this kind are the eight relations $\mathsf{DC}, \mathsf{EC}, \mathsf{PO}, \mathsf{EQ}, \mathsf{TPP}, \mathsf{NTPP}, \mathsf{TPP}^{-1}, \mathsf{NTPP}^{-1}$. In the best known approach in this domain, the Region Connection Calculus by Randell, Cui, and Cohn [138], these relations are known as the RCC-8 relations. The Region Connection Calculus and in particular the RCC-8 relations are the central topic of this work and will be described in detail in the following chapter.

What distinguishes the different approaches and what thereby influences the definable relations is the interpretation of the connectedness relation and the properties of the considered regions. Some approaches distinguish between open and closed regions [134, 5] which allows, for instance, to define different kinds of contact. Asher and Vieu [5] distinguished between strong contact (two regions have points in common) and weak contact (two regions are disjoint but their topological closures share common points). Other approaches do not make this distinction and treat regions which are open, closed, or neither equally [138]. The Region Connection Calculus [138] considers only the topological closure of regions: two regions are connected if their topological closures share a common point. Cohn et al. [28] argue that this definition is more appropriate for commonsense spatial reasoning since there is "no reason to believe that some physical objects occupy closed regions and others open". Orthogonal to the interpretation of the connectedness relation is the distinction of what regions are considered. A very common restriction is using only non-empty regular regions. As shown by Asher and Vieu [5], models based on Clarke's connectedness relation require all   regions to be nonempty and regular. However, it is possible to specify additional properties of regions such as dimensionality, internal connectedness, i.e., whether a region consists of one-piece or of multiple pieces, or the existence of holes. The different approaches are compared in [29, 30].

All of these approaches have in common that the relations are axiomatized and defined in first-order logic which provides them with formal semantics. The formal properties of first-order theories based on a connectedness relation were studied by Grzegorczyk [78], Dornheim [39] and Pratt and Schoop [132, 148]. It turns out that all of them are incomplete and, moreover, reasoning about these theories is undecidable. However, this changes when considering *constraint languages* over relations definable from the connectedness relation. These can be regarded as the special case of first-order sentences where only existentially quantified region variables are used. Bennett [10] gave an encoding of the RCC-8 relations in propositional modal logic and, thus proved that reasoning over the RCC-8 constraint language is decidable.

A different approach to defining topological relations was given by Egen-hofer [44] in the area of spatial information systems. Egenhofer defined binary relations according to the 9 different intersections of the interior, exterior, and boundary of regions, hence, called *9-intersection*. Depending on the regions that are used, many different relations can be defined in this way [45, 46]. If only two-dimensional, internally connected regular regions without holes are considered and only emptiness or non-emptiness of the intersection is taken into account, this results in the same set of eight base relations as definable in the above described approaches. In contrast to the connection based approaches, this approach is not provided with formal semantics which makes it very difficult to study its formal properties. For instance, it was tried to identify sound and complete algorithms for reasoning over the eight relations defined by Egenhofer [151, 48] while it was taken for granted that path-consistency decides consistency if only base relations are used. As shown by [77], this is not the case for Egenhofer's definition of the eight topological relations and it is still an open problem whether reasoning over these relations is decidable. A more detailed description of Egenhofer's approach and the connections to RCC-8 is given in Section 4.4.

3.3.2 Orientation

Orientation is, like topology, very well suited for a qualitative approach. In everyday (non-technical) communication, orientation of spatial entities with respect to other spatial entities is usually given in terms of a qualitative category like "to the left of" or "northeast of" rather than using a numerical expression like "53 degrees" (which is certainly more common in technical communication like in aviation). Unlike the topological approaches we discussed in the previous section, orientation of spatial entities is a ternary relationship depending on the located object, the reference object, and the *frame of reference* which can be specified either by a third object or by a given direction. In the literature one distinguishes between three different kinds of frames of reference, extrinsic ("external factors impose an orientation on the reference object"), intrinsic ("the orientation is given by some inherent property of the reference object"), and deictic ("the orientation is imposed by the point of view from which the reference object is seen") [87, p.45]. If the frame of reference is given, orientation can be expressed in terms of binary relationships with respect to the given frame of reference.

Most approaches to qualitatively dealing with orientation are based on points as the basic spatial entities and consider only two-dimensional space. Frank [56] suggested different methods for describing the cardinal direction of a point with respect to a reference point in a geographic space, i.e., directions are in the form of "north", "east", "south", and "west" depending on the granularity. These are, however, just labels which can be equally termed as, for instance, "front", "right", "back", and "left" in a local space. Frank distinguishes between the "cone-based" method and the "projection-based"

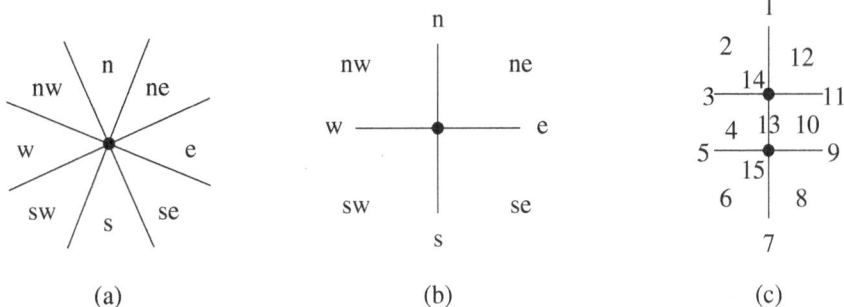

Fig. 3.1. Orientation relations between points: (a) cone-based (b) projection-based (c) double-cross

method for determining the different sectors corresponding to the single directions (see Figure 3.1). The projection-based approach allows to represent the nine different relations (n, ne, e, se, s, sw, w, nw, eq) in terms of the point algebra (see Section 2.5) by specifying a point algebraic relation for each of the two axes separately. This provides the projection-based approach (which is also called the "cardinal algebra" [116]) with formal semantics which were used by Ligozat [116] to study its computational properties. In particular, Ligozat found that reasoning with the cardinal algebra is NP-complete and, further, identified a maximal tractable subset of the cardinal algebra by using the concept of "preconvex" relations, a method which has already been used for Allen's interval algebra [115].

A further point-based approach was developed by Freksa [58], the so-called "double-cross" calculus, which defines the direction of a located point to a reference point with respect to a perspective point. Within this approach three axes are used, one is specified by the perspective point and the reference point, the other two axes are orthogonal to the first one and are specified by the reference point and the perspective point, respectively. These axes define 15 different ternary base relations (see Figure 3.1c). The computational properties of this calculus have only recently been studied [149]. It turned out that reasoning with the 15 base relations alone is already NP-hard.

Developing orientation relations between extended spatial entities is much more difficult than between points. Extended objects often have their own intrinsic directions like a natural front. Also, the direction between extended objects with complex shapes such that, for instance, their convex hulls intersect is not at all clear. Even for simpler objects there is often no agreement about which natural language expression describes their orientational relationship best. Therefore, it is mostly tried to use approximations to the spatial regions or use only spatial regions of a particular kind. One approach which has been chosen by many researchers results in restricting all regions to be rectangles whose sides are parallel to the axes determined by the frame of reference [79, 129, 6]. In this approach all regions can be represented by their

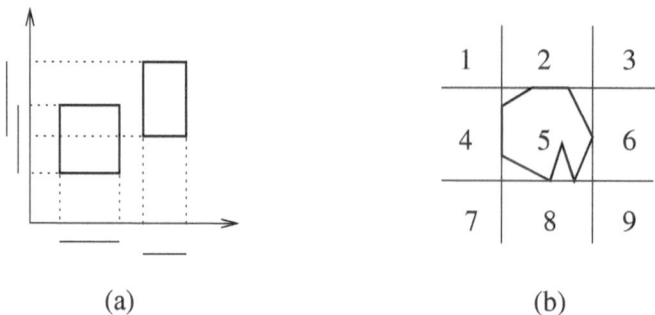

(a)   (b)

Fig. 3.2. Orientation relations between extended entities: (a) rectangle algebra (b) direction-relation matrix

projections to the defining axes which corresponds to having Allen's interval algebra for each axis separately, i.e., every relation is a pair of two interval relations (see Figure 3.2a). For two-dimensional space this results in 13×13 different base relations (also called the "rectangle algebra" [6]) whose formal semantics are provided by the interval algebra.

Balbiani et al. [6, 7] studied the formal properties of the rectangle algebra (and also of the "block algebra" which is the n-dimensional extension of the interval algebra [8]). NP-completeness of the algebra carries over from the interval algebra. Balbiani et al. [6, 7] identified a tractable subset of the rectangle algebra following a line of reasoning which has been introduced by Ligozat [115], namely, by considering convex and preconvex relations. Unlike for the interval algebra, the set of preconvex relations is not closed under the fundamental operations. Thus, Balbiani et al. extended the concept of preconvexity by distinguishing between weakly and strongly preconvex relations, and show that the set of strongly preconvex relations is a tractable subset of the rectangle algebra for which path-consistency is sufficient for deciding consistency. In fact the rectangle algebra represents more than just orientation between two rectangles but also their topological relations. Hence, the rectangle algebra can be regarded as an approach to combining topology and orientation. However, because of the large number of relations of the rectangle algebra (a total number of 2^{169} relations) reasoning even over a tractable subset can be very inefficient.

An interesting but less expressive approach to representing orientational relationships between extended spatial entities was presented by Goyal and Egenhofer [76]. They introduced a 3×3 "direction-relation matrix" which represents the 9 sectors formed by the minimal bounding axes of an extended spatial entity (see Figure 3.2b). For each sector it is possible to specify whether the located object is contained in the sector or not, or (nonqualitatively) to which degree the located object is contained in the sector. Goyal and Egenhofer proposed a reasoning procedure for the qualitative version, but because formal semantics have not been specified it is not clear whether the composition of relations other than base relations was correctly defined.

3.3.3 Distance

Together with topology and orientation, distance is one of the most important aspects of space. Unlike the other two, distance is a scalar entity. Dealing with distance information is an important cognitive ability in our everyday life. In order to grab something, for instance, we must be good in judging distances. When communicating about distances, we usually use qualitative categories like "A is close to B" or qualitative distance comparatives like "A is closer to B than to C", but also numerical values like "A is about one meter away from B". As indicated by the above examples, one can distinguish between absolute distance relations (the distance between two spatial entities) and relative distance relations (the distance between two spatial entities as compared to the distance to a third entity). While absolute distance can be represented either qualitatively or quantitatively, relative distance is purely qualitative. When representing absolute distance in a qualitative way, this also depends on the scale of space which is used. Montello [124] suggests four different kinds of scales of space, figural space, vista space, environmental space, and geographic space.

Most approaches to qualitative distance consider points as the basic entities. Absolute distance relations are obtained, e.g., by dividing the real line into several sectors such as "very close", "close", "commensurate", "far", and "very far" depending on the chosen level of granularity [88]. Relative distance can be obtained by comparing the distance to a given reference distance which results in ternary relations such as "closer than", "equidistant", or "farther than". Reasoning about qualitative distances leads to several difficulties. For instance, given a sequence of aligned points $p_1, \ldots, p_n$ such that p_i is close to p_{i+1} for every i, for which n is p_n far from p_1? Moreover, combining distance relations does not only depend on the distances itself but also on the position of the corresponding points. For instance if point B is far from A and C is far from B, then C can be very far from A if A, B, and C are aligned and if B is between A and C; or C can be close to A if the angle between AB and BC is small. Therefore, it seems advisable to study distance in combination with orientation. This combination is called *positional* information.

One approach for developing a position calculus is by Clementini et al. [24] who combine a cone-based orientation approach with absolute distance relations (see Figure 3.3a). Clementini et al. present different procedures for computing the composition of two positional relations (A, B) and (B, C). They consider three special cases where BC is the same, opposite, or orthogonal direction to AB. Another approach is by Isli et al. [97] who propose several position calculi on various levels of granularity by combining relative distance relations with different approaches to orientation such as the

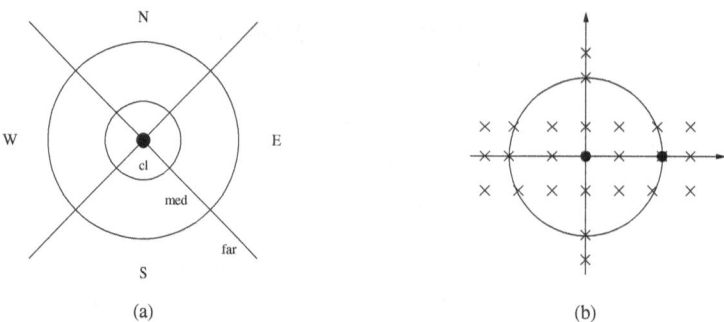

Fig. 3.3. Different approaches to representing positional information: (a) absolute distances combined with cone-based orientation [24] (b) relative distances combined with projection-based orientation [97]

projection-based approach (see Figure 3.3b) or the double-cross calculus. The computational properties of these approaches have not been studied yet.

4. The Region Connection Calculus

The Region Connection Calculus (RCC) by Randell, Cui, and Cohn [138] and in particular RCC-8 is probably the best known approach to qualitative spatial reasoning. This is documented by the citations of RCC which appear in almost any paper on qualitative spatial reasoning and by the numerous research papers written on RCC itself. In this chapter we give a detailed introduction to the Region Connection Calculus and especially to RCC-8 which is the central topic of this book. A more detailed overview of the work on the Region Connection Calculus can be found in [28].

In the following section we summarize the original definitions and axioms of the Region Connection Calculus in first-order logic as given by Randell et al. [138]. Hence, the title of this section is the same as of Randell et al.'s paper. In Section 4.2 we introduce RCC-8 and give an encoding of RCC-8 in modal logic in Section 4.3 which is a slight modification of Bennett's original encoding [12]. In Section 4.4 we introduce Egenhofer's system of topological relations and compare it to the Region Connection Calculus.

4.1 A Spatial Logic Based on Regions and Connection

The Region Connection Calculus (RCC) developed by Randell, Cui, and Cohn [138] is a topological approach to spatial representation and reasoning where *spatial regions* are non-empty regular subsets of some topological space $\mathcal{U}$. Spatial regions do not have to be *internally connected*, i.e., they might consist of (multiple) disconnected pieces. Since all spatial regions are regular subsets of the same topological space $\mathcal{U}$, all spatial regions have the same dimension, namely, the dimension of $\mathcal{U}$ (provided that $\mathcal{U}$ has a particular dimension).

RCC is based on a single primitive relation between spatial regions, the "connected" relation C. The intended topological interpretation of $C(a, b)$, where a and b are spatial regions, is that a and b are connected if and only if their topological closures share a common point. Within this interpretation it is not distinguished between open, semi-open, and closed regions which is different from previous approaches by Randell and Cohn [135, 134] and Clarke [23, 22]. The only requirements of the relation C is that it is reflexive and symmetric which is enforced by the following two axioms:

J. Renz: Qualitative Spatial Reasoning with Topological Information, LNAI 2293, pp. 41–50, 2002.
© Springer-Verlag Berlin Heidelberg 2002

$$\forall x C(x, x) \tag{4.1}$$

$$\forall x, y[C(x, y) \rightarrow C(y, x)] \tag{4.2}$$

Using $C(x, y)$, a large number of different relations can be defined. Among those are the following relations, their meaning under the intended interpretation of the C relation is given in brackets [138]: $P(x, y)$ (x is a part of y), $PP(x, y)$ (x is a proper part of y), $EQ(x, y)$ (x is equal to y), $O(x, y)$ (x overlaps y), $PO(x, y)$ (x partially overlaps y), $DR(x, y)$ (x is discrete from y), $EC(x, y)$ (x is externally connected with y), $TPP(x, y)$ (x is a tangential proper part of y), $NTPP(x, y)$ (x is a non-tangential proper part of y). The relations P, PP, TPP, and $NTPP$ are non-symmetrical, their converses are denoted by $P^{-1}, PP^{-1}, TPP^{-1}$, and $NTPP^{-1}$, respectively. The formal definition of these relations is the following [138]:

$$DC(x, y) \equiv_{def} \neg C(x, y) \tag{4.3}$$

$$P(x, y) \equiv_{def} \forall z[C(z, x) \rightarrow C(z, y)] \tag{4.4}$$

$$PP(x, y) \equiv_{def} P(x, y) \wedge \neg P(y, x) \tag{4.5}$$

$$EQ(x, y) \equiv_{def} P(x, y) \wedge P(y, x) \tag{4.6}$$

$$O(x, y) \equiv_{def} \exists z[P(z, x) \wedge P(z, y)] \tag{4.7}$$

$$PO(x, y) \equiv_{def} O(x, y) \wedge \neg P(x, y) \wedge \neg P(y, x) \tag{4.8}$$

$$DR(x, y) \equiv_{def} \neg O(x, y) \tag{4.9}$$

$$EC(x, y) \equiv_{def} C(x, y) \wedge \neg O(x, y) \tag{4.10}$$

$$TPP(x, y) \equiv_{def} PP(x, y) \wedge \exists z[EC(z, x) \wedge EC(z, y)] \tag{4.11}$$

$$NTPP(x, y) \equiv_{def} PP(x, y) \wedge \neg\exists z[EC(z, x) \wedge EC(z, y)] \tag{4.12}$$

$$P^{-1}(x, y) \equiv_{def} P(y, x) \tag{4.13}$$

$$PP^{-1}(x, y) \equiv_{def} PP(y, x) \tag{4.14}$$

$$TPP^{-1}(x, y) \equiv_{def} TPP(y, x) \tag{4.15}$$

$$NTPP^{-1}(x, y) \equiv_{def} NTPP(y, x) \tag{4.16}$$

Using these relations, it is also possible to define Boolean functions such as $sum(x,y)$ (the union of x and y), $compl(x)$ (the complement of x), $prod(x,y)$ (the intersection of x and y), and $diff(x,y)$ (the difference of x and y). The functions $compl, prod$, and $diff$ are partial since their result might be the empty region which is undefined. These functions can be used to define other relations such as internal connectedness of regions:

$$CON(x) \equiv_{def} \forall y\forall z[EQ(sum(y, z), x) \rightarrow C(y, z)] \tag{4.17}$$

Additional axioms can be used to specify properties of spatial regions. The following axiom states that every region has a non-tangential proper part, i.e., there are no atomic regions:

$$\forall x\exists y NTPP(y, x) \tag{4.18}$$

Gotts [72] showed that every regular connected T_3 space is a model for the RCC axioms, i.e., every regular subset of such a topological space fulfills the axioms.

The relations, functions, and axioms presented here are just a small fraction of what can be expressed within the RCC theory. For instance, it is possible to add other primitive relations and functions such as the convex-hull function [138, 35]. Gotts [71, 73] studied a large number of different relations which can be defined upon the C relation. Of particular interest are those relations that form a set of jointly exhaustive and pairwise disjoint base relations. If these relations are closed under composition they generate a relation algebra, thus, reasoning about these relations can be done using the methods described in Section 2.4. What is needed is mainly a composition table which can be computed using the first-order definitions of the RCC relations. Depending on the level of granularity, many different sets of base relations can be defined within the RCC theory.

Randell et al. [138] suggested a set of eight base relations, later denoted as RCC-8. This set of relations is interesting for a number of reasons. It is the smallest set of base relations which allows topological distinctions rather than just mereological (being expressible by using the part-whole relationship). Most other relations definable in the RCC theory are refinements of these relations. As such, RCC-8 is ideally suited as a starting point for qualitative spatial reasoning which can be extended in many different ways. Furthermore, the semantics of these relations can be described by using propositional logics rather than first-order logics [10, 12], a property which allows us to prove decidability.

4.2 The Region Connection Calculus RCC-8

The Region Connection Calculus RCC-8 is the constraint language formed by the eight jointly exhaustive and pairwise disjoint base relations DC, EC, PO, EQ, TPP, NTPP, TPP^{-1}, and $NTPP^{-1}$ definable in the RCC-theory and by all possible unions of the base relations. It can be easily verified by the first-order definitions given in the previous section that exactly one of the eight base relations holds between any two spatial regions. Unions of possible base relations are used to represent indefinite knowledge. Since the base relations are pairwise disjoint, this results in $2^8 = 256$ different RCC-8 relations altogether (including the empty relation and the universal relation). In some papers the set of base relations is denoted as RCC-8 while the set of all possible unions of base relations is denoted as 2^{RCC8}. We will, however, use RCC-8 to refer to the set of all possible disjunctions of the base relations and $\mathcal{B}$ to refer to the set of base relations. Analogous to the general RCC-theory, spatial regions in RCC-8 are non-empty regular subsets of some topological space that do not have to be internally connected, and do not have a particular dimension. Without loss of generality (due to the intended interpretation

Table 4.1. Topological interpretation of the eight base relations of RCC-8. All spatial regions are regular closed, i.e., $x = c(i(x))$ and $y = c(i(y))$. $i(\cdot)$ specifies the topological interior of a spatial region, $c(\cdot)$ the topological closure

RCC-8 *Relation*	*Topological Constraints*
$DC(x, y)$	$x \cap y = \emptyset$
$EC(x, y)$	$i(x) \cap i(y) = \emptyset,\ x \cap y \neq \emptyset$
$PO(x, y)$	$i(x) \cap i(y) \neq \emptyset,\ x \not\subset y,\ y \not\subset x$
$TPP(x, y)$	$x \subset y,\ x \not\subset i(y)$
$TPP^{-1}(x, y)$	$y \subset x,\ y \not\subset i(x)$
$NTPP(x, y)$	$x \subset i(y)$
$NTPP^{-1}(x, y)$	$y \subset i(x)$
$EQ(x, y)$	$x = y$

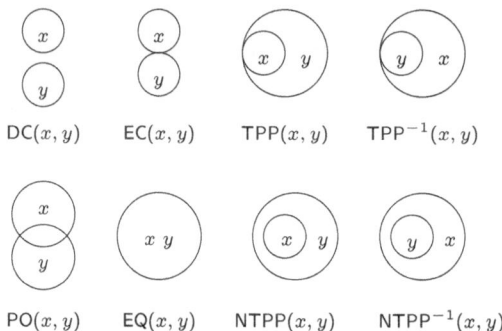

Fig. 4.1. Two-dimensional examples for the eight base relations of RCC-8

of the C relation) we require spatial regions to be regular *closed* subsets of a topological space.

The RCC-8 relations can be given a straightforward topological interpretation in terms of point-set topology (see Table 4.1), which is almost the same as for the topological relations given by Egenhofer [44] (though Egenhofer places stronger constraints on the domain of regions, e.g., regions must be one-piece and are not allowed to have holes, see Section 4.4). Examples for the RCC-8 base relations are given in Figure 4.1.

Converse, intersection and union of relations can easily be obtained by performing the corresponding set theoretic operations. Composition of base relations can be computed using the formal definitions of the relations given in the previous section [137, 10]. The compositions of the eight base relations are shown in Table 4.2. Every entry in the composition table specifies the relation obtained by composing the base relation of the corresponding row with the base relation of the corresponding column. Composition of two disjunctive RCC-8 relations can be obtained by computing the union of the composition of the base relations. Note that the composition table corresponds to the ex-

Table 4.2. Composition table for the eight base relations of RCC-8, where $*$ specifies the universal relation

$\circ$	DC	EC	PO	TPP	NTPP	TPP^{-1}	NTPP^{-1}	EQ
DC	$*$	DC,EC PO,TPP NTPP	DC,EC PO,TPP NTPP	DC,EC PO,TPP NTPP	DC,EC PO,TPP NTPP	DC	DC	DC
EC	DC,EC PO,TPP^{-1} NTPP^{-1}	DC,EC PO,TPP TPP^{-1},EQ	DC,EC PO,TPP NTPP	EC,PO TPP NTPP	PO TPP NTPP	DC,EC	DC	EC
PO	DC,EC PO,TPP^{-1} NTPP^{-1}	DC,EC PO,TPP^{-1} NTPP^{-1}	$*$	PO TPP NTPP	PO TPP NTPP	DC,EC PO,TPP^{-1} NTPP^{-1}	DC,EC PO,TPP^{-1} NTPP^{-1}	PO
TPP	DC	DC,EC	DC,EC PO,TPP NTPP	TPP NTPP	NTPP	DC,EC PO,TPP TPP^{-1},EQ	DC,EC PO,TPP^{-1} NTPP^{-1}	TPP
NTPP	DC	DC	DC,EC PO,TPP NTPP	NTPP	NTPP	DC,EC PO,TPP NTPP	$*$	NTPP
TPP^{-1}	DC,EC PO,TPP^{-1} NTPP^{-1}	EC,PO TPP^{-1} NTPP^{-1}	PO TPP^{-1} NTPP^{-1}	PO,EQ TPP TPP^{-1}	PO TPP NTPP	TPP^{-1} NTPP^{-1}	NTPP^{-1}	TPP^{-1}
NTPP^{-1}	DC,EC PO,TPP^{-1} NTPP^{-1}	PO TPP^{-1} NTPP^{-1}	PO TPP^{-1} NTPP^{-1}	PO TPP^{-1} NTPP^{-1}	PO,TPP^{-1} TPP,NTPP NTPP^{-1},EQ	NTPP^{-1}	NTPP^{-1}	NTPP^{-1}
EQ	DC	EC	PO	TPP	NTPP	TPP^{-1}	NTPP^{-1}	EQ

tensional definition of composition given in Section 2.4.1 only if the universal region is not permitted [15].

A spatial configuration can be described by specifying a set Θ of constraints over RCC-8, written as xRy or $R(x,y)$, where R is an RCC-8 relation and x, y are *spatial variables* over the infinite domain of all possible spatial regions. An important reasoning problem is deciding consistency of Θ, i.e., deciding whether there is an assignment of non-empty, regular closed regions of some topological space to variables of Θ in a way that all constraints are satisfied. We call this problem RSAT, or RSAT($\mathcal{S}$) if only relations of a specific set $\mathcal{S}$ are used in Θ. RSAT is a sub-problem of CSPSAT which is defined in Section 2.4.2 and which can be tackled using the same methods, for instance, enforcing path-consistency as a partial method for deciding consistency. Other useful reasoning problems include the minimal labels problem RMIN and the entailment problem RENT, the sub-problems of CSPMIN and CSPENT, respectively.

Another set of jointly exhaustive and pairwise disjoint base relations definable in the RCC-theory on a coarser level of granularity than RCC-8 is RCC-5 [10]. For RCC-5 the boundary of a region is not taken into account, i.e., one does not distinguish between DC and EC and between TPP and NTPP. These relations are combined to the RCC-5 base relations DR and PP, respectively (see Section 4.1). Thus, RCC-5 contains the five base relations DR, PO, PP, PP^{-1}, and EQ and all 2^5 possible disjunctions thereof. The RCC-5 relations are closed under composition and form a relation algebra. The composition table for RCC-5 is given in Table 4.3. In this work we will focus on RCC-8,

Table 4.3. Composition table for the five base relations of RCC-5

∘	DR	PO	PP	PP^{-1}	EQ
DR	*	DR,PO,PP	DR,PO,PP	DR	DR
PO	DR,PO,PP^{-1}	*	PO,PP	DR,PO,PP^{-1}	PO
PP	DR	DR,PO,PP	PP	*	PP
PP^{-1}	DR,PO,PP^{-1}	PO,PP^{-1}	PO,PP,PP^{-1},EQ	PP^{-1}	PP^{-1}
EQ	DR	PO	PP	PP^{-1}	EQ

but most of our results can easily be applied to RCC-5.

4.3 Encoding of RCC-8 in Modal Logic

Another way of solving problems concerning RCC-8 is using the encoding of the relations in first order logic. Such an encoding does not lead to efficient decision procedures, however. In order to overcome this problem, Bennett [10, 12] used different encodings of RCC-8 in propositional intuitionistic and modal logic. In this work we will use a slight modification of Bennett's encoding of RCC-8 in propositional modal logic [12]. An introduction to modal logics is given in Section 2.2.2.

Remark 4.1. Throughout this work we will use the following convention for referring to spatial regions, spatial variables, and propositional atoms corresponding to spatial regions or spatial variables:

- Spatial variables are written as x, y, z.
- Spatial regions are written as X, Y, Z.
- Propositional atoms corresponding to spatial regions or spatial variables are written as X, Y, Z.

If the same letter is used in different fonts in the same context, it represents the same region. For instance, X is a possible instance of x, Y a possible instance of y, and X is the propositional atom corresponding to x or to X.

Bennett obtained the modal encoding by analyzing the relationship of regions to the universe $\mathcal{U}$. For the modal encoding we are using, Bennett restricted his analysis to closed regions that are connected if they share a point and overlap if they share an interior point.[1] If, e.g, X and Y are disconnected,

---

[1] There is also a modal encoding based on open regions which is not as simple as the encoding based on closed regions [12].

Table 4.4. Bennett's encoding of the eight base relations in modal logic [12]

Relation	Model Constraints	Entailment Constraints
$DC(x, y)$	$\neg(X \wedge Y)$	$\neg X, \neg Y$
$EC(x, y)$	$\neg(IX \wedge IY)$	$\neg(X \wedge Y), \neg X, \neg Y$
$PO(x, y)$	—	$\neg(IX \wedge IY), X \rightarrow Y, Y \rightarrow X, \neg X, \neg Y$
$TPP(x, y)$	$X \rightarrow Y$	$X \rightarrow IY, Y \rightarrow X, \neg X, \neg Y$
$TPP^{-1}(x, y)$	$Y \rightarrow X$	$Y \rightarrow IX, X \rightarrow Y, \neg X, \neg Y$
$NTPP(x, y)$	$X \rightarrow IY$	$Y \rightarrow X, \neg X, \neg Y$
$NTPP^{-1}(x, y)$	$Y \rightarrow IX$	$X \rightarrow Y, \neg X, \neg Y$
$EQ(x, y)$	$X \rightarrow Y, Y \rightarrow X$	$\neg X, \neg Y$

the complement of the intersection of X and Y is equal to the universe. Further, both regions must not be empty, i.e., the complements of both X and Y are not equal to the universe. In the same way all topological constraints corresponding to the RCC-8 relations (see Table 4.1) can be written as constraints of the form $(m = \mathcal{U})$ and $(e \neq \mathcal{U})$, where m and e are set-theoretic expressions, denoted as *model constraints* and *entailment constraints*, respectively [10]. In the above example, $\overline{X \cap Y}$ is the model constraint and $\overline{X}$ and $\overline{Y}$ are the entailment constraints. Any model constraint must hold, whereas no entailment constraint must hold [10].

For some of the constraints it is necessary to refer to the interior of regions. For this purpose the topological interior operator i is used. This operator must satisfy the following constraints for arbitrary sets $\Phi, \Psi \subseteq \mathcal{U}$ [12]:

$$i(\Phi) \quad \subseteq \quad \Phi, \tag{4.19}$$

$$i(i(\Phi)) \quad = \quad i(\Phi), \tag{4.20}$$

$$i(\mathcal{U}) \quad = \quad \mathcal{U}, \tag{4.21}$$

$$i(\Phi \cap \Psi) \quad = \quad i(\Phi) \cap i(\Psi). \tag{4.22}$$

The model and entailment constraints can be encoded in modal logic, where regions correspond to propositional atoms, the interior operator i corresponds to a modal operator $\mathbf{I}$ (see Table 4.4), and the universe $\mathcal{U}$ corresponds to the set of all worlds W [12]. The constraints for i must also hold for the modal operator $\mathbf{I}$, which results in the following axiom schemata [12] for arbitrary modal formulas ϕ, ψ:

$$\mathbf{I}\phi \quad \rightarrow \quad \phi, \tag{4.23}$$

$$\mathbf{II}\phi \quad \leftrightarrow \quad \mathbf{I}\phi, \tag{4.24}$$

$$\mathbf{I}\top \quad \leftrightarrow \quad \top \text{ (for any tautology } \top), \tag{4.25}$$

$$\mathbf{I}(\phi \wedge \psi) \quad \leftrightarrow \quad \mathbf{I}\phi \wedge \mathbf{I}\psi. \tag{4.26}$$

Axiom schemata 4.23 and 4.24 correspond to the modal axioms **T** and **4** and axiom schemata 4.26 and 4.26 already hold for any modal logic K, so **I** is a modal S4-operator (see Section 2.2.2).

The four axiom schemata specified by Bennett are not sufficient to exclude non-closed regions as it was intended. In order to account for that, we add one formula for each atom X, which corresponds to the topological property of regular closed regions:

A regular closed region is the closure of an open region. $\neg X$ specifies the complement of X, and, thus, $\neg I \neg X$ the closure of X.

$$X \leftrightarrow \neg I \neg I X \tag{4.27}$$

Note that the S4 encoding can be used to reason about any kind of open or closed regions. Both the non-emptiness constraint, i.e., the entailment constraint $\neg X$, and the regularity constraint (4.27) are optional and can be regarded as properties of regions definable in the modal representation. They are needed to make the representation conform to the intended interpretation of the original RCC theory.

In order to combine the different model and entailment constraints, Bennett [12] uses another modal operator $\Box$. $\Box \phi$ is interpreted as $\phi = \mathcal{U}$ and $\neg \Box \phi$ as $\phi \neq \mathcal{U}$. Since m is a model constraint if $m = \mathcal{U}$ holds, any model constraint m can be written as $\Box m$ and any entailment constraint e as $\neg \Box e$. If $\Box X$ is true in a world w of a model $\mathcal{M}$, written as $\mathcal{M}, w \Vdash \Box X$, then X must be true in any world of $\mathcal{M}$. So $\Box$ is an S5-operator with the constraint that all worlds are mutually accessible. Therefore Bennett calls it a *strong S5-operator* [12]. Now all model and all entailment constraints containing the strong S5-operator can be conjunctively combined to a single modal formula. So the modal encoding of RCC-8 is made with an S4-operator that corresponds to the topological interior operator and a strong S5-operator that is used to obtain a single modal formula.

4.4 Egenhofer's Approach to Topological Spatial Relations

Independently of the Region Connection Calculus a different approach to topological spatial relationship was developed by Egenhofer in the area of geographical information systems (GIS). Egenhofer [44] classified the relationships between two spatial entities according to the nine possible intersections of their interior, exterior, and boundary, hence, called the *9-intersection-model*. Depending on the nature of the considered spatial entities, many different relationships can be expressed by this model. For instance, it is possible to use spatial entities of different dimensions or distinguish different degrees of intersection. In his first approach, however, Egenhofer restricted the domain of spatial entities to be two-dimensional spatial regions whose boundary is a closed Jordan curve, i.e., simply connected planar regions that are not allowed to have holes. When looking at only whether the nine intersections for this kind of spatial regions are empty or non-empty and when eliminating all impossible relations, this results in eight different binary topological

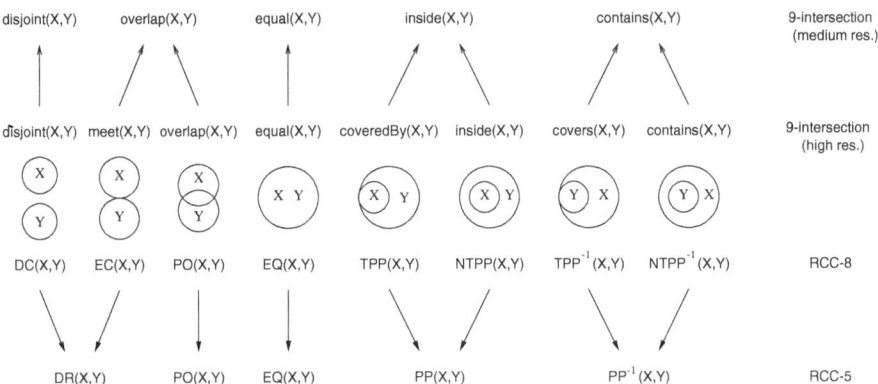

Fig. 4.2. Comparison of different systems of topological relations: the Region Connection Calculi RCC-8 and RCC-5 and the high and medium resolution of Egenhofer's 9-intersection model

relationships. The Region Connection Calculus and the 9-intersection model, which are two completely different approaches to topological relationships, lead to exactly (apart from the different constraints on regions) the same set of topological relations. Thus, there seems to be a natural agreement about what is a reasonable level of granularity of topological relations.

The computational properties of Egenhofer's topological relations were studied by Grigni et al. [77] who considered two different notions of satisfiability, the purely syntactical notion of *relational consistency* and the semantic notion of *realizability*, which are both different from what we call consistency. Relational consistency means that there is a path-consistent refinement of all relations to base relations, realizability means that there is a model consisting of simply connected planar regions. Orthogonal to this distinction, Grigni et al. [77] considered different sets of JEPD relations. One of them is the original set of eight base relations suggested by Egenhofer which they called the *high resolution* case. Another set consists of five base relations, the *medium resolution* case, which are obtained by combining some of the high resolution relations. Similar to RCC-5, the medium resolution relations do not distinguish relationships according to the boundary of regions. The difference of this set to RCC-5 is that the relations EC and PO are combined to form a new base relation, whereas for RCC-5 the relations DC and EC are combined (see Figure 4.2). With this distinction, the medium resolution relations can also be used to represent the possible relationships between non-topological sets: overlap(a, b) means that the sets a and b have a non-empty intersection while none of them is a subset of the other, if disjoint(a, b) holds, the sets a and b have no elements in common. This set-theoretic interpretation is not possible for RCC-5, since it distinguishes between interior and boundary elements of the sets. If two sets share some boundary elements, they are still in the DR relationship.

By reducing the NP-hard string-graph problem [105, 106], Grigni et al. [77] showed that deciding realizability is NP-hard for the high and medium resolution cases even if only constraints over the base relations are used. It is an open problem whether the realizability problem is in NP and even whether it is decidable. The reduction of the string graph problem, however, is possible only because all regions must be simply connected planar regions. Hence, this result does not carry over to the consistency problem of RCC-8 where spatial regions can be of any dimension and internal connectedness is not required. The relational consistency problem, which is obviously tractable if only base relations are used, was shown to be NP-hard for the high and medium resolution cases if all disjunctions over the base relations are permitted.

5. Cognitive Properties of Topological Spatial Relations

It has always been a motivation for qualitative reasoning in general and qualitative spatial reasoning in particular that the ways of representing and reasoning about knowledge in these fields are considered to be similar to the ways of human cognition. In many cases, however, this purely relies on the researchers intuition rather than on actual empirical data. As Cohn [26, p.22] wrote in his overview article on qualitative spatial reasoning, "An issue which has not been much addressed yet in the QSR literature is the issue of cognitive validity – claims are often made that qualitative reasoning is akin to human reasoning, but with little or no empirical justification." This is particularly true for the topological calculi described in Chapter 4 which are studied throughout this work. Although Randell et al. [138] and Egenhofer [44] identified the same set of eight topological relations using completely different approaches, and there thus seems to be a natural agreement about what is a reasonable level of granularity of topological relations, there is no indication so far that these relations are also interesting from a cognitive point of view.

In this chapter, two empirical investigations are presented that were conducted in collaboration with Markus Knauff and Reinhold Rauh from the Center of Cognitive Science of the Albert-Ludwigs University of Freiburg where we tried to evaluate the cognitive adequacy (this term will be defined below) of topological relations. Both investigations rely on the grouping task as an empirical method. In our first study we present subjects a set of items each with two different colored circles in different spatial relationships that have to be grouped according to their similarity. Since it is possible to group the items according to different kinds of spatial relationships such as size, distance, orientation, or topology of the two circles, we wanted to find out whether topological relationship is an important grouping criterion and at which level of granularity the topological relations are used. In our second study we weakened constraints on the shape of the presented regions and used regular polygons instead of circles. These regions offer a larger number of possible spatial relationships than circles, especially different ways of how two regions can touch each other. The goal of this second study was to see whether and how this influences subjects' groupings.

J. Renz: Qualitative Spatial Reasoning with Topological Information, LNAI 2293, pp. 51–64, 2002.
© Springer-Verlag Berlin Heidelberg 2002

5.1 Psychological Background

It is often claimed in the literature that some approach is cognitively adequate or cognitively valid. However, there seems to be no general agreement on what *cognitive adequacy* actually means. In a strong sense cognitive adequacy must at least mean that something is a model of human cognition. Strube [153] proposed to range it from an absolutely strong meaning down to a very weak notion of conforming to well known ergonomic standards. The first case means that a formal approach or implementation is claimed to be an adequate model of human knowledge and reasoning mechanisms in all or some relevant aspects. In the second case the claim is reduced to the assumption that an implementation of a formal approach can be characterized as ergonomic and user-friendly [153].

Orthogonal to this view, Knauff et al. [103] suggested that another distinction is very important as well. In their view, determining the cognitive adequacy of a qualitative spatial reasoning formalism has to be broken down to at least two sub-aspects, namely the representational and the inferential aspect of the calculus. Knauff et al. [103] termed these two sub-concepts of cognitive adequacy *conceptual adequacy* and *inferential adequacy*. Conceptual adequacy refers to the degree of how a set of entities and relations correspond to the "mental vocabulary" of a person. Inferential adequacy refers to the degree of how the computational inference algorithm (e.g. constraint satisfaction, theorem proving, ...) conforms to the human mental reasoning process. A cognitive adequate qualitative reasoning system would therefore be based on a set of spatial relations that humans also use quite naturally, and would draw inferences in a way people would do with respect to certain criteria (type of inferences; order of inferences; preferred, easy and hard inferences; in the extreme maybe also in accordance with processing latencies). The determination of the cognitive adequacy would be of interest for basic research in the field of the psychology of knowledge and the psychology of thinking on the one hand, and for Computer Science and Artificial Intelligence on the other. But mostly, applied research like cognitive ergonomics and human-computer-interaction (HCI) research would benefit from these results, since these could justify the decision for one candidate out of a set of competing spatial calculi in spatial information systems. In our investigations we concentrated on the problem of conceptual adequacy of spatial relations and its empirical determination.

Looking for empirical methods to determine the conceptual adequacy, many results can be found in psychological research on conceptual structure. But most theories are about the conceptual structure of entities like natural kind terms or artifacts (see for example the overviews given in Komatsu [104], and Rips [146]). Relational concepts, however, were not investigated very heavily, and therefore psychological methods are tuned to the assessment of the conceptual structure of entities, but not for relations. Other research on spatial relations comes from (psycho-)linguistics where mostly acceptability ratings of natural language expressions like "above", "below", "left", "right" and so on, were obtained (see for example [85, 162]). These results are heavily dependent on the investigated language and may, therefore, not reveal everything about people's conceptualization of space; this is due to the fact that different languages may provide different means to express the conceptualizations, and linguistic expressions are differently "cut" to cover spatial configurations, as one can easily verify when comparing English prepositions with those of other languages, even with those of close relatives like German. Therefore, it is important to use also language independent methods to determine the human conceptualization of space. Mark et al. [120] mentioned some of them, like the graphic examples task, the graphic comparison task, and the grouping task. Especially, the *grouping task* seems to be one prominent empirical method that is easy to carry out and to communicate to the participant. The subject is given a sample of spatial configurations that she/he is prompted to group together. The basic idea is that the conceptualization of space guides the grouping of items into categories. These observable categories give important hints on the subjects' conceptualization and may serve to exclude certain sets of relations from being cognitively adequate. Subsequent intensional descriptions of the groupings may provide additional hints (1) what informational content was used by the subject to group the items, and (2) to lower the risk of choosing the wrong intension of extension-equivalent categories.

5.2 Empirical Investigation I: Grouping Task with Circular Regions

The goal of our first empirical evaluation was to find first evidences of whether the systems of topological relations developed by Egenhofer [44] and by Randell et al. [138] are cognitively adequate and which level of granularity is more adequate. For this investigation we start with very simple regions, namely, circles that can be related in only a limited number of ways. If it turns out that topological relations are important for grouping these simple regions, we can extend our investigation and use more complex regions. Otherwise, topological relations would not seem to be interesting from a cognitive point of view.

5.2.1 Subjects, Method, and Procedure

20 students (10 female, 10 male) of the University of Freiburg had to accomplish a grouping task that consisted of 96 items showing a configuration of a red and a blue circle. Figure 5.1 shows some examples of these items. After solving some practice trials to get acquainted with the procedure they then saw the color screen as depicted in Figure 5.2. The 96 items of spatial

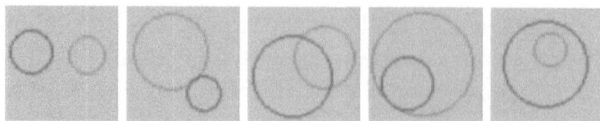

Fig. 5.1. Items with two different colored circles as used in our first empirical investigation

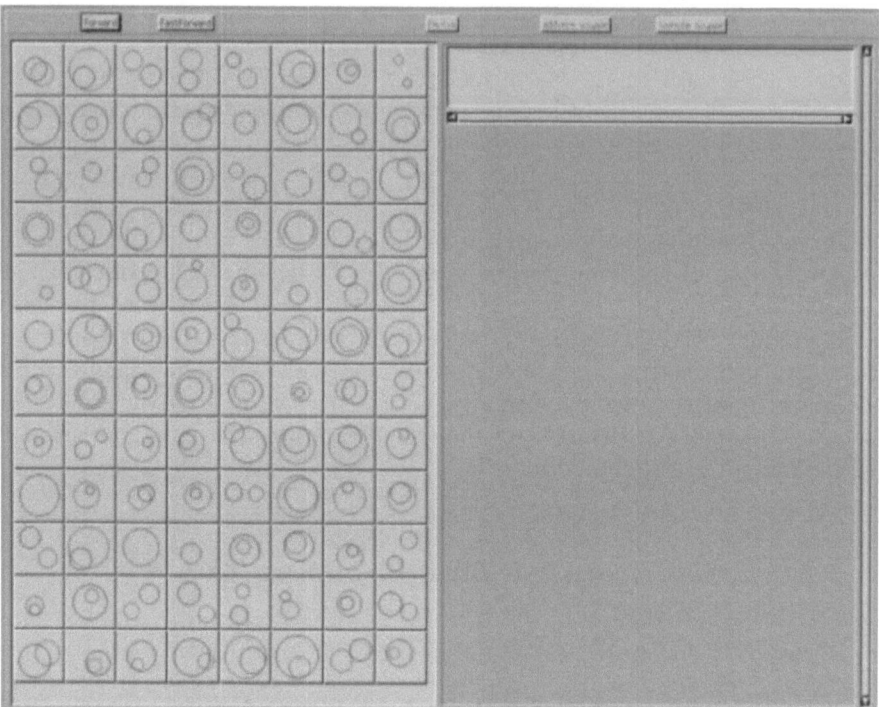

Fig. 5.2. Screen-shot of the monitor at the beginning of the first grouping task. Subjects had to move the pictures on the left into groups at the right side of the screen. The two buttons at the right gave the possibility to add as many groups as needed and to remove unused groups

configurations of the red and blue circle consisted of 14 DC, 11 EC, 12 PO, 10 EQ, 15 TPP, 11 NTPP, 11 TPP^{-1}, and 12 NTPP^{-1} instances. The number of occurrences was chosen in a random fashion around the number 12±3 in order to avoid a bias towards RCC-8 grouping by using equally frequent groups. All subjects had to judge the same items because the aggregation of data had to be done across subjects per item. Instances for each relation were generated by a program that produced instances of the RCC-8 relations randomly with respect to metrical and ordering information.

After the grouping phase of the empirical investigation, subjects had to describe their groupings by natural language description in German. We did this to obtain further information on the criteria that guided subjects' grouping behavior, and we applied this (unexpected) phase at the end of the empirical session in order to avoid influences of verbalizations on the pure grouping behavior itself.

5.2.2 Results of the First Investigation

Subjects needed from 14 to 35 minutes with an average of 20 minutes to group items and to describe their groupings. Aggregating grouping answers over all 20 subjects, we obtained a 96×96 matrix of Hamming distances between items that were the basis of a subsequent cluster analysis. Figure 5.3 shows the dendrogram—combining items and clusters first that are nearest to each other from left to right—of the cluster analysis with the method of average linkage between groups. In order to avoid interpreting clustering results that are only the consequence of different clustering methods, we computed cluster analyses with different clustering methods, but could not find any differences to the results as obtained by the above mentioned method—clustering results were invariant across different clustering methods.

The first important result is that after a few clustering steps, all instances were clustered according to the RCC-8 relations. Therefore, we left out the "dendrites" of single instances and only display clustering on higher levels to test for compatibility of the empirical results with the RCC-5 and medium resolution case of Egenhofer's conceptualization. As it can be easily seen, there is no evidence that items were grouped together according to one of these sets of relations; so our empirical results indicate that RCC-5 and the medium resolution set of Egenhofer's relations are not conceptually adequate.

The clustering of TPP and NTPP with their inverses can be attributed to disregarding the relationship of a reference object with a to-be-localized object, and that subjects judged the configuration as a whole where one object had another object as a tangential proper part or non-tangential proper part, respectively.

Furthermore, the verbalizations of subjects' groupings were analyzed by categorizing them whether they included purely topological (T), orientational (O), and size information (S), or combinations of them. In Table 5.1 we present the results of the categorization of subjects' verbalizations, as completed by two raters and showing nearly perfect concordance (97.18%). Table 5.1 clearly shows that the informational content of most verbalizations was due to topological information alone or in combination with other information (62.1% + 14.1% + 19.2% of 390 judgments, i.e. 95.4%.) The size information used in the verbalizations was purely qualitative.

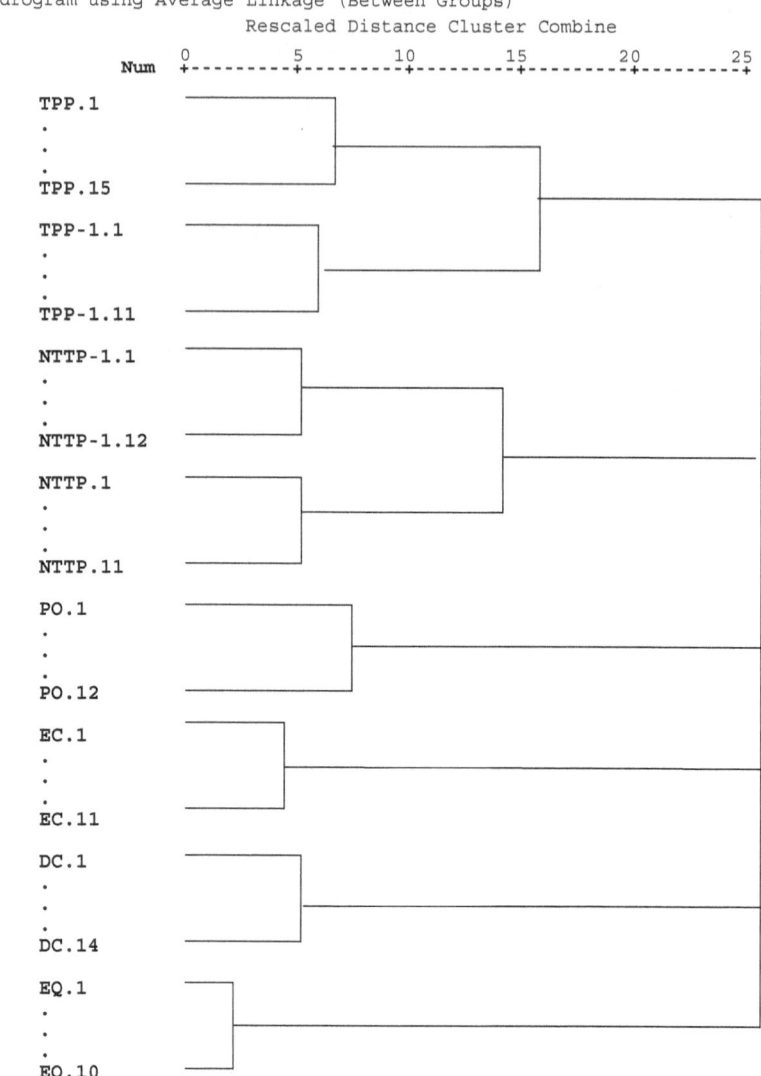

Fig. 5.3. Dendrogram of the cluster analysis of 96 items of the first investigation based on Hamming distances between items

Table 5.1. Categorization of subjects' verbalizations of their groupings of the first investigation (n=390 verbalizations, 195 by each rater)

Informational content	Percentage
Topology (T)	62.1%
Orientation (O)	–
Size (S)	–
T + O	14.1%
T + S	19.2%
O + S	–
T + O + S	–
Other	4.6%

5.2.3 Discussion

What we have found in our first empirical grouping task is that (i) topological relationships are a relevant and dominating factor in judging spatial configurations, (ii) that the topological relations as specified in RCC-8 seem to form a relevant level of conceptualizing spatial configurations, and (iii) that neither RCC-5 nor the medium resolution variant of Egenhofer's relations can be found on a coarser level of granularity. Even if it becomes clear that the two five relation sets are for the most part cognitively irrelevant, it is an interesting observation that items of the relations DC and EC were grouped together more often than EC and PO. The last observation was (iv) that exploratory results seem to indicate that orientation information and qualitative size information were also considered in grouping spatial configurations. This is a relevant point in our results because one important task in qualitative spatial reasoning is to extend topological calculi for other non-topological concepts like distance, direction, or size aspects. If one wants to make this extension in such a way that the resulting calculi can still be called cognitively adequate, our empirical results provide useful hints in which direction to extend topological calculi.

In general, the results indicated that the RCC-8 relations or Egenhofer's definition of topological relations, respectively, are actually the most promising starting point for further psychological investigations on human conceptual topological knowledge. However, further evidence in particular investigations with more complex regions than circles will be needed before RCC-8 can be claimed as cognitively adequate.

5.3 Empirical Investigation II: Grouping Task with Polygonal Regions

In our second investigation we weaken the constraints on the shape of the regions. In the first investigation all regions were circles (see Figure 5.1), whereas in our second investigation the regions are regular polygons such as squares, pentagons or hexagons. The main difference to using circles is that when circles externally connect each other or are tangential proper part of each other, their boundaries always touch each other at a single point, whereas regular polygons with these properties can touch each other at one or more points, at a line, or even at both. Some of the different possibilities of how regular polygons can touch each other are shown in Figure 5.4. Note

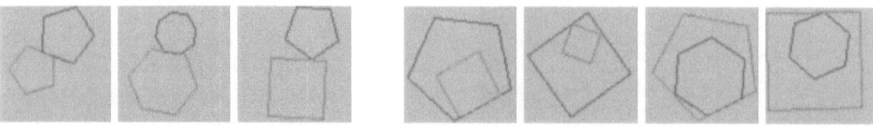

Fig. 5.4. Different possibilities of external and internal connection of regular polygons

that in this investigation we refer to the Region Connection Calculus when referring to topological relations. This is only for increasing readability. Since all regions consist of one piece and have no holes, all statements we make about the RCC-8 relations in this section are equally true for the eight relations defined by Egenhofer.

5.3.1 Subjects, Method, and Procedure

19 students (10 female, 9 male) of the University of Freiburg participated for course credit. Subjects had to accomplish a grouping task (on a Sun Workstation) that consisted of 200 items showing a configuration of a red and a blue regular polygon as shown in Figure 5.4. For the sample of 200 items we randomly selected 25 ± 5 items for each of the eight RCC-8 relations with the following restrictions. We generated regular polygons with four to eight corners, and made sure that for each possibility of how two regions can connect each other there were at least some items.

After solving some practice trials to get acquainted with the procedure subjects saw the color screen as depicted in Figure 5.5. The screen was split into two different sections. The left section contained the items and the right section the groupings subjects had to form. The items had to be moved with a mouse from the left to the right. In order not to bias the number of groupings, in the beginning only one group was displayed on the right, and subjects had the possibility to add as many groupings as they wanted. The 200 items in the left section were arbitrarily split into four different windows such that subjects could switch between the four windows at any time.

Similar to our first investigation, all subjects had to judge the same 200 items because the aggregation of data had to be done across subjects per item. After the grouping phase, subjects had to describe their groupings by natural language descriptions in German. For each subject the complete procedure

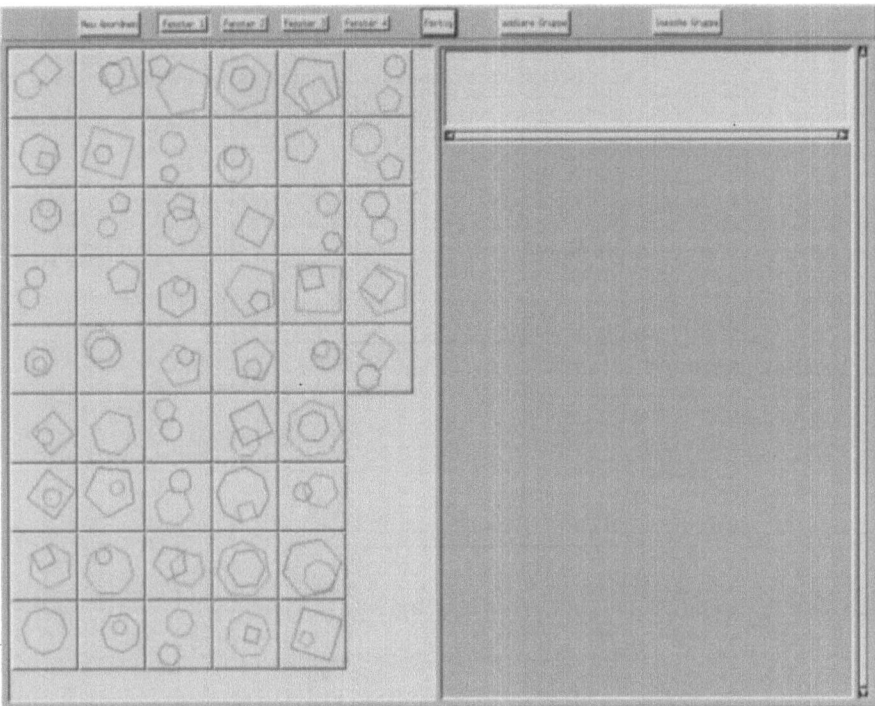

Fig. 5.5. Screen-shot of the monitor at the beginning of the second grouping task

was stored in a log-file and we wrote a tool which we could use to trace the complete procedure afterwards.

5.3.2 Results of the Second Investigation

Subjects needed about 45 minutes to group items and to describe their groupings. By aggregating grouping answers over all 19 subjects, we obtained a 200×200 matrix of Hamming distances between items that was the basis of a cluster analysis using the method of average linkage between groups. Unlike the results of our first investigation where only some subjects did not distinguish between TPP and TPP^{-1} and between NTPP and NTPP^{-1}, this time the result is more unique. As it can be seen in the dendrogram of Figure 5.6, there is no distinction between NTPP and NTPP^{-1} and between TPP and TPP^{-1} at all. As in our first study, the clustering of TPP and NTPP with their converses can be attributed to disregarding the relationship of a reference object with a to-be-localized object. However, there are sub-clusters of items of TPP or TPP^{-1} and of items of EC. By looking at the items of the different sub-clusters, we found that these sub-clusters belong exactly to the different ways two regular polygons can be externally or internally connected

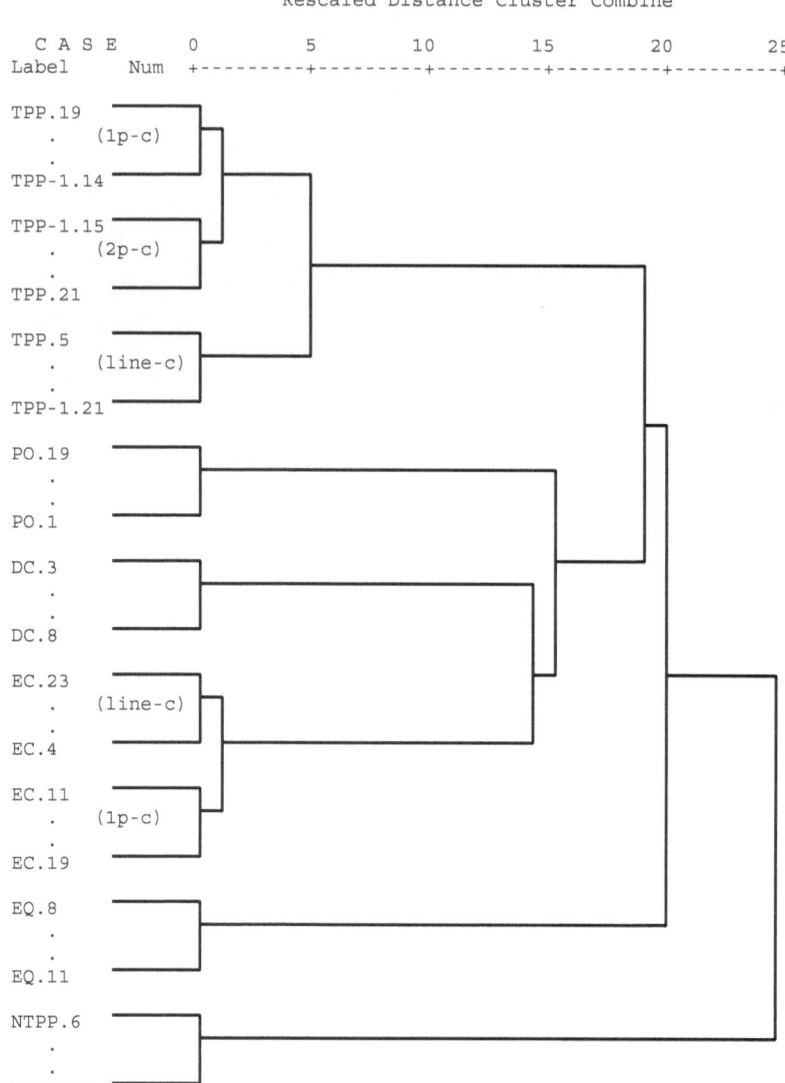

Fig. 5.6. Dendrogram of the cluster analysis of 200 items of the second investigation based on Hamming distances between items. Note that the labels given on the left are only for identification purposes. The three sub-clusters of the TPP and TPP^{-1} items have the property of one-point contact (1p-c), two-point contact (2p-c), or line-contact (line-c), respectively. For the EC items there are two sub-clusters of items with line contact and one-point contact

to each other, either at a single point, at two points, or at a line. The clustering of groups on a higher level of the cluster analysis is due to idiosyncratic categorizations or mis-groupings of single items by a few subjects (one subject wrote that he used one group as a clipboard and forgot to assign these items in the end).

Furthermore, the verbalizations of subjects' groupings were analyzed by categorizing them whether they included purely topological (T), orientational (O), and size information (S). Additionally, we introduced two new categories for the degree of contact of the boundaries (C) and mentioning of pure shape. Additionally, we also considered all possible combinations of these five factors. In Table 5.2, we present the results of the categorization of subjects' verbalizations, as completed by one independent rater.

Table 5.2. Categorization of subjects' verbalizations of their groupings of the second investigation (n=164 verbalizations)

Informational content	Percentage
Topology (T)	42.1%
Orientation (O)	–
Size (S)	–
Contact (C)	1.8%
Shape	–
T + O	14.6%
T + C	36.6%
T + S	1.2%
Other	3.4%

Table 5.2 clearly shows that the informational content of nearly all verbalizations incorporated topological information alone or in combination with other information ($42.1\% + 14.6\% + 36.6\% + 1.2\% = 94.5\%$). Size and orientational information was only used in combination with topological information. The category "other" in Table 5.2 consists of verbalizations that do not belong to any combination of T, O, M, C, or S, but were mostly misinterpretations of the EQ items. The most surprising result, however, is that a substantial number of verbalizations mentioned the degree of contact as an additional modifier of topological relations. This confirms also the findings of the cluster analysis, where EC, TPP and TPP^{-1} configurations had sub-clusters that differed with respect to the degree of contact.

This observation was confirmed by evaluating the log-files we collected during the experiments. The "trace" tool enabled us to recover the final screen the subjects produced when finishing the grouping and the verbalization tasks (see Figure 5.7). We found that except for one subject, who grouped the items belonging to TPP and NTPP and to TPP^{-1} and NTPP^{-1} together and who distinguished items showing disconnected regions according to their direction, all other subjects did not distinguish the color of the regions presented on the objects, i.e., they did not distinguish between NTPP and NTPP^{-1} and between TPP and TPP^{-1}. Table 5.3 shows the detailed evaluation of the final screens of the subjects. By RCC-8* we denote the set of RCC-8 relations without distinction of the reference object and the to-be-localized object. Figure 5.7 gives an example of the most common result, where the subject distinguished different kinds of external and internal connection.

Table 5.3. Evaluation of the subjects' final screen

#subjects	description of their groupings
1	TPP and NTPP grouped together, DC plus direction
5	RCC-8*
1	RCC-8* plus directions
1	RCC-8* plus size of overlapping area
1	RCC-8* plus different connections for TPP
6	RCC-8* plus different connections for EC and TPP
3	RCC-8* plus different connections for EC and TPP and PO
1	RCC-8* plus different connections for EC and TPP, and different directions for PO and DC

5.3.3 Discussion

For our second investigation, we increased the different possibilities of how subjects can group items by partially lifting the constraints on the shape of the regions we used. Except for one subject, all others grouped the items according to the distinctions made by the RCC-8 relations. Even those subjects who took also orientation or size into account made this only to further refine the RCC-8 relations but never used orientation or size independently of the topological relations. This result gives further evidence that topological relations in general are an important part of human spatial cognition, and that the level of granularity given by the RCC-8 relations is particularly important. Topological distinctions on a coarser level of granularity do not seem to be cognitively adequate. This confirms the observations made in our first investigation.

Many subjects also distinguished items according to a finer level of granularity than given by the RCC-8 relations. 11 out of 19 subjects distinguished between the different possibilities of how two regions can touch each other from the outside or from the inside. These distinctions can all be made on a purely topological level without taking into account other spatial aspects such as direction, distance, or shape. Thus, it appears that topological relationships are much more important in human spatial cognition than any other relationships between spatial regions.

Although almost all subjects did not distinguish between TPP and TPP^{-1} and between NTPP and NTPP^{-1}, we do not consider the set of six relations

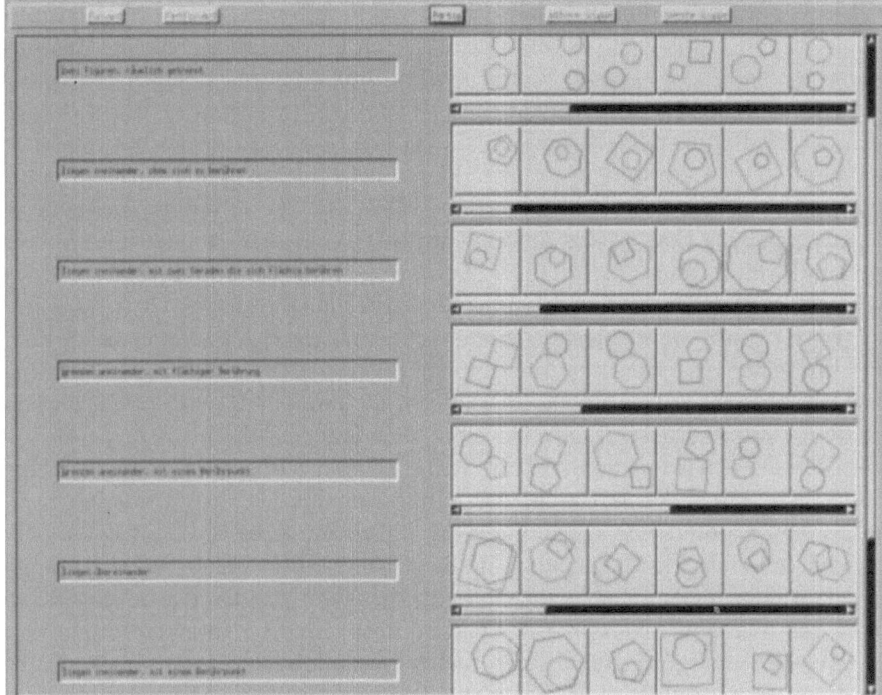

Fig. 5.7. Screen-shot of the recovered final screen of one of the subjects

we called RCC-8* as cognitively adequate. As already explained, this is due to disregarding the relationship of a reference object with a to-be-localized object. We believe that when this distinction is explicitly emphasized in the instructions, subjects will group items accordingly. Furthermore, it can be derived from our results in Chapter 6 that reasoning with RCC-8* base relations is NP-hard whereas reasoning with RCC-8 base relations is tractable.

5.4 Discussion and Outlook

Qualitative reasoning is often considered to be akin to human reasoning. In many cases this relies only on the researchers intuition. Whether or not a formal approach to spatial relations is a cognitively adequate model of human spatial knowledge is in fact an empirical question and can be answered only on the basis of empirical results. The goal of our empirical investigations was to find out whether the topological relations used in qualitative spatial reasoning can be regarded as cognitively adequate and which level of granularity seems to be more adequate.

Both investigations reported above suggest that topological relations are an important part of human spatial knowledge. With a high level of agreement subjects classified items showing two regions in different spatial relationships according to topological distinctions. In both investigations the distinctions made by the RCC-8 relations were the most important ones, which gives clear evidences of the cognitive adequacy of the RCC-8 relations. Relations on a coarser level of granularity such as the RCC-5 relations did not appear to be adequate. From a cognitive point of view, these results give additional motivation for studying the RCC-8 relations. Thus, for developing a more general theory of space that covers different aspects such as distance, direction, or size and which can be regarded as cognitively adequate, topological relations and particularly the RCC-8 relations appear to be an excellent starting point.

Future work on cognitive adequacy of topological relations includes further lifting the restrictions on the shape of the presented regions. One question is whether subjects continue to distinguish the number of connections and up to which number. In our investigation some subjects distinguished between connections at a single point, at two points, and at a line. It will be interesting to see whether subjects distinguish connections at $1, 2, 3, \ldots, n$ points, and more than n points and which number n turns out to be the most commonly chosen (maybe it is the magical number 7 ± 2, the capacity of the human short term memory [122]). For arbitrary shapes of regions it is also possible that two regions connect at two or more lines. Another interesting question is how subjects group items when regions are allowed to consist of multiple pieces or to have holes. Answering this question also allows to judge the adequacy of the different restrictions on regions as given by Egenhofer [44] and by Randell et al. [138].

So far we have only investigated conceptual cognitive adequacy of systems of topological relations. It will be interesting to see how humans reason about these relations, and to investigate the inferential cognitive adequacy of the reasoning methods used in qualitative spatial reasoning (cf. [103]).

6. Computational Properties of RCC-8

In the previous chapter we found evidence that topological relations in general and RCC-8 in particular are a fundamental part of human conceptualization of space. This makes the RCC-8 relations very interesting as a basis for our goal of obtaining an expressive spatial calculus covering several aspects of space which is both cognitively adequate and offers efficient reasoning mechanisms. While the first condition, cognitive adequacy, is fulfilled to some degree (as shown in the previous chapter), little is known about the computational properties of RCC-8–neither whether efficient reasoning mechanisms for RCC-8 exist nor whether efficient reasoning over RCC-8 is possible at all (cf. [77]).

In this chapter we study the computational properties of RCC-8 by analyzing a fundamental reasoning problem of RCC-8, the problem of deciding consistency of a set of constraints over RCC-8, called RSAT. Most other reasoning problems of RCC-8 can be polynomially reduced to RSAT. Unfortunately, the consistency problem of RCC-8 turns out to be NP-hard, i.e., provided that P $\neq$ NP there are no complete reasoning mechanisms which decide all instances of RSAT in polynomial time. Instead of designing efficient approximation algorithms, we choose to pursue a strategy which has also been used in the area of qualitative temporal reasoning, namely, to identify tractable fragments of RSAT, i.e., subsets of RCC-8 for which the consistency problem can be decided in polynomial time. Those fragments that contain all base relations can then be used in a divide-and-conquer-like manner to obtain complete algorithms which solve the consistency problem more efficiently. The goal of such an analysis is to find complete algorithms which solve almost all RSAT instances up to a certain size efficiently.

In order to identify a tractable fragment of RCC-8, we use a modification of Bennett's [12] encoding of RCC-8 in modal logic (see Section 4.3) for transforming RSAT to propositional satisfiability (SAT). Those relations that are transformed to propositional Horn formulas, which is a tractable fragment of SAT, form a tractable subset of RCC-8. It turns out that this set of RCC-8 relation, denoted $\widehat{\mathcal{H}}_8$, is also maximal with respect to tractability, i.e., if any other relation is added to $\widehat{\mathcal{H}}_8$, deciding consistency of constraints over the new set becomes NP-hard. Having identified a maximal tractable subset of RCC-8, we are also interested in whether the most popular partial method for

J. Renz: Qualitative Spatial Reasoning with Topological Information, LNAI 2293, pp. 65–116, 2002.
© Springer-Verlag Berlin Heidelberg 2002

deciding consistency, the path-consistency method, is complete for deciding consistency of $\widehat{\mathcal{H}}_8$. This is important since the path-consistency method is very simple and runs in time $O(n^3)$. By relating positive unit resolution, a complete method for deciding satisfiability of propositional Horn formulas, to the path-consistency method, we are able to prove that path-consistency decides consistency for sets of constraints over $\widehat{\mathcal{H}}_8$. Using the same technique, we identify an efficient algorithm for finding a consistent scenario of every consistent set Θ of constraints over $\widehat{\mathcal{H}}_8$, i.e., a refinement Θ' of Θ for which all constraints are assertions over the RCC-8 base relations.

6.1 Computational Complexity of RCC-8

In this section we prove that reasoning with RCC-8 is NP-hard by showing that reasoning with a subset of RCC-8 is already NP-hard. We will then show how complexity results for subsets of RCC-8 can be carried over to other subsets of RCC-8, and, using this result, give NP-hardness proofs for a number of different subsets $\mathcal{S}$ of RCC-8. All of these proofs use a transformation from a propositional satisfiability problem to $\mathsf{RSAT}(\mathcal{S})$ by constructing a set of spatial constraints Θ for every instance $\mathcal{I}$ of the propositional satisfiability problem, such that Θ is consistent if and only if $\mathcal{I}$ is a positive instance. The propositional satisfiability problems we use are 3SAT, the problem of deciding whether there is a truth assignment for a set of clauses where each clause has exactly three literals, as well as two variants of 3SAT where truth assignments of particular types are required. These variants are NOT-ALL-EQUAL-3SAT, the problem of deciding whether there is a truth assignment such that for every clause at least one literal is assigned *true* and one literal is assigned *false*, and ONE-IN-THREE-3SAT, the problem of deciding whether there is a truth assignment such that for every clause exactly one literal in every clause is assigned *true*. All three decision problems are NP-hard [147].

The different transformations we use in this section as well as in Section 6.3 have in common that every variable v of the propositional satisfiability problem is transformed to two RCC-8-constraints $x_v \{R_t, R_f\} y_v$ and $x_{\neg v} \{R_t, R_f\} y_{\neg v}$ corresponding to the positive and the negative literal of v, where R_t and R_f are RCC-8-relations with $R_t \cap R_f = \emptyset$. v is assigned *true* if and only if $x_v \{R_t\} y_v$ holds and assigned *false* if and only if $x_v \{R_f\} y_v$ holds. Since the two literals corresponding to a variable need to have opposite assignments, we have to make sure that $x_v \{R_t\} y_v$ holds if and only if $x_{\neg v} \{R_f\} y_{\neg v}$ holds, and *vice versa*, for which additional "polarity constraints" are required. In addition, every literal occurrence l of the propositional satisfiability problem is transformed to the RCC-8-constraint $x_l \{R_t, R_f\} y_l$, where $x_l \{R_t\} y_l$ holds if and only if l is assigned *true*. In order to assure the correct assignment of positive and negative literal occurrences with respect to the corresponding variable, polarity constraints are required again. For instance, if the variable v is assigned *true*, i.e., $x_v \{R_t\} y_v$ holds, then $x_p \{R_t\} y_p$ must

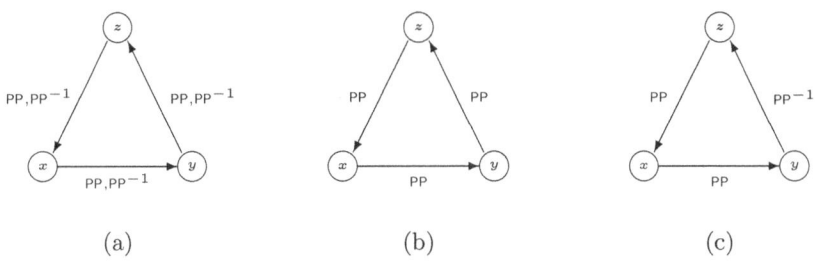

Fig. 6.1. Property 6.1: (a) is the original configuration, (b) is an inconsistent, (c) is a consistent refinement of the relations to base relations

hold for every positive literal occurrence p of v, and $x_n\{R_f\}y_n$ must hold for every negative literal occurrence n of v. Further, "clause constraints" have to be added to assure that the clause requirements of the specific propositional satisfiability problem are satisfied. For example, if $\{i, j, k\}$ is a clause of an instance of **ONE-IN-THREE-3SAT**, then exactly one of the constraints $x_i\{R_t\}y_i$, $x_j\{R_t\}y_j$, and $x_k\{R_t\}y_k$ must hold.

In the following we will prove NP-hardness for RCC-5, since the result can be immediately transferred to RCC-8. For this proof we will use **NOT-ALL-EQUAL-3SAT** and set R_t to $\{PP\}$ and R_f to $\{PP^{-1}\}$. The polarity and the clause constraints of the transformation are based on the following two properties that can be verified using the composition table (Table 4.2).

Property 6.1. *Let x, y, z be spatial variables, where the relation $\{PP, PP^{-1}\}$ holds between all of them. Then any refinement of these relations to base relations such that a path in the constraint graph starting and ending at x, and passing y and z contains only $\{PP\}$ or only $\{PP^{-1}\}$ is inconsistent (see Figure 6.1)[1].*

Property 6.2. *Let x, y, z be spatial variables, where $x\{PO\}y$, $z\{PP, PP^{-1}\}x$, and $z\{PP, PP^{-1}\}y$ hold. Then a refinement of these relations to base relations is only consistent, if the relations between (z, x) and between (z, y) are refined to the same base relation (see Figure 6.2).*

With Property 6.1 it can be ruled out that all constraints are refined to the same base relation (not-all-equal). Using Property 6.2 a refinement of one constraint can be propagated to another constraint.

Theorem 6.3. RSAT(RCC-5) *is NP-hard.*

---

[1] In the constraint graphs displayed in the figures of this section, spatial variables are symbolized as circles, the relation $\{PO\}$ is symbolized as a dotted line. All other relations are symbolized with a labeled arrow such that $x \xrightarrow{R} y$ symbolizes the constraint xRy.

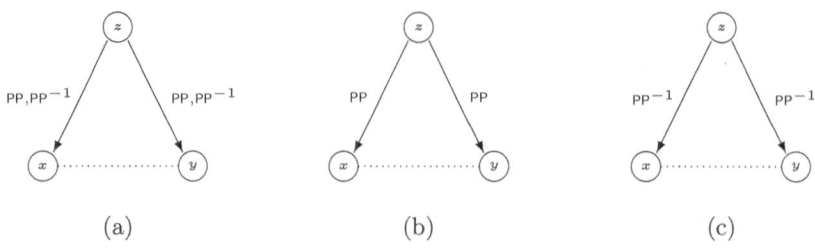

Fig. 6.2. Property 6.2: (a) is the original configuration, (b) and (c) are the only consistent refinements of the relations to base relations

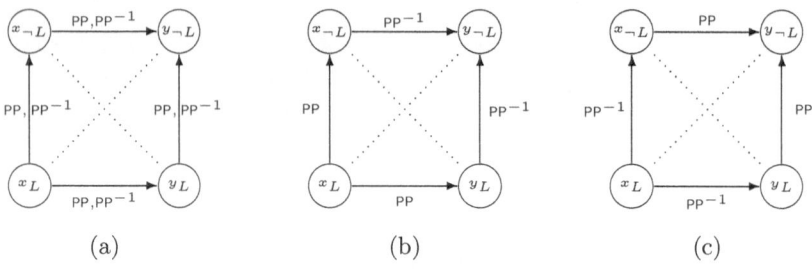

Fig. 6.3. The polarity constraints (a) for the transformation of NOT-ALL-EQUAL-3SAT assure that positive and negative literals of the same variable have opposite assignments: (b) and (c) are the only consistent refinements of the relations to base relations

Proof. Transformation of NOT-ALL-EQUAL-3SAT to RSAT(RCC-5) (see also Grigni et al [77]). Let $\mathcal{V} = \{v_1, v_2, \ldots, v_n\}$ be a set of variables and $\mathcal{C} = \{c_1, c_2, \ldots, c_m\}$ be a set of clauses of an arbitrary instance of NOT-ALL-EQUAL-3SAT with $c_i = \{l_{i,1}, l_{i,2}, l_{i,3}\}$, where $l_{i,j}$ are literal occurrences over variables of $\mathcal{V}$. We will construct a set of spatial constraints Θ, such that Θ is satisfiable if and only if $\mathcal{C}$ is a positive instance of NOT-ALL-EQUAL-3SAT using the following three transformation steps:

(1) For each variable $v_L \in \mathcal{V}$ the spatial variables $x_L, y_L, x_{\neg L}$ and $y_{\neg L}$ are introduced by adding the spatial constraints $x_L\{PP, PP^{-1}\}y_L$ and $x_{\neg L}\{PP, PP^{-1}\}y_{\neg L}$ to Θ. Additionally, the following polarity constraints are added to Θ (see Figure 6.3a):

$$x_L\{PP, PP^{-1}\}x_{\neg L}, y_L\{PP, PP^{-1}\}y_{\neg L},$$
$$x_L\{PO\}y_{\neg L}, y_L\{PO\}x_{\neg L}.$$

(2) For each literal occurrence $l_{i,j}$ the spatial variables $x_{i,j}$ and $y_{i,j}$ are introduced by adding the spatial constraint $x_{i,j}\{PP, PP^{-1}\}y_{i,j}$ to Θ. Depen-

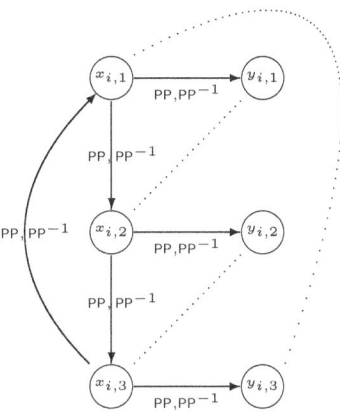

Fig. 6.4. Transformation of a not-all-equal clause $c_i = \{l_{i,1}, l_{i,2}, l_{i,3}\}$ to spatial constraints

ding on whether the literal occurrence is positive or negative, different polarity constraints have to be added to Θ.

a) $l_{i,j} \equiv v_L$:

$$x_{i,j}\{\mathsf{PP}, \mathsf{PP}^{-1}\}x_{\neg L}, y_{i,j}\{\mathsf{PP}, \mathsf{PP}^{-1}\}y_{\neg L},$$
$$x_{i,j}\{\mathsf{PO}\}y_{\neg L}, y_{i,j}\{\mathsf{PO}\}x_{\neg L}.$$

b) $l_{i,j} \equiv \neg v_L$:

$$x_{i,j}\{\mathsf{PP}, \mathsf{PP}^{-1}\}x_L, y_{i,j}\{\mathsf{PP}, \mathsf{PP}^{-1}\}y_L,$$
$$x_{i,j}\{\mathsf{PO}\}y_L, y_{i,j}\{\mathsf{PO}\}x_L.$$

(3) For each clause $c_i = \{l_{i,1}, l_{i,2}, l_{i,3}\}$ the following clause constraints are added to Θ (see Figure 6.4):

$$x_{i,1}\{\mathsf{PP}, \mathsf{PP}^{-1}\}x_{i,2}, x_{i,2}\{\mathsf{PP}, \mathsf{PP}^{-1}\}x_{i,3}, x_{i,3}\{\mathsf{PP}, \mathsf{PP}^{-1}\}x_{i,1},$$
$$x_{i,1}\{\mathsf{PO}\}y_{i,3}, x_{i,2}\{\mathsf{PO}\}y_{i,1}, x_{i,2}\{\mathsf{PO}\}y_{i,3}.$$

With this transformation for every literal as well as for every literal occurrence two spatial variables x and y (with the appropriate indices) are introduced. When a literal occurrence or a literal is assigned *true*, the corresponding spatial variables hold the relation $x\{\mathsf{PP}\}y$, when a literal occurrence or a literal is assigned *false*, the corresponding spatial variables hold the relation $x\{\mathsf{PP}^{-1}\}y$.

Transformation step (1) introduces the spatial variables corresponding to the positive and the negative literal of each variable. Property 6.2 assures that positive and negative literals have opposite assignments. This is shown in Figure 6.3. Transformation step (2) introduces spatial variables for every literal occurrence. Again, Property 6.2 assures correct assignments. Finally, transformation step (3) together with Property 6.1 makes sure that the not-all-equal condition of the literals of every clause is also fulfilled by the corresponding spatial variables. We now have to show that an instance of NOT-ALL-EQUAL-3SAT has a solution if and only if the set of spatial constraints Θ obtained by the given transformation is consistent.

RSAT$\Rightarrow$ NOT-ALL-EQUAL-3SAT: Suppose that the set of spatial constraints Θ obtained by transformation from a given instance Σ of NOT-ALL-EQUAL-3SAT is consistent, and suppose that θ is a consistent instantiation of Θ. Then an assignment σ that satisfies Σ can be obtained in the following way: For every variable $v_L \in \mathcal{V}$, if $\theta(x_L)\{$PP$\}\theta(y_L)$ holds, then $\sigma(v_L)$ is *true*, otherwise $\sigma(v_L)$ is *false*.

NOT-ALL-EQUAL-3SAT $\Rightarrow$ RSAT: Suppose that Σ is a positive instance of NOT-ALL-EQUAL-3SAT, and suppose that σ is an assignment that satisfies Σ. Then the set of spatial constraints Θ obtained by transformation from Σ with respect to σ is consistent. We will show this by constructing a spatial configuration that satisfies all relations of Θ. Before that we will point out some properties of Θ. Let Θ' be the set of spatial constraints transformed from Σ with transformation steps (1) and (2) with respect to σ, i.e., if, for example, $\sigma(v_L) = true$ then $x_L\{$PP$\}y_L$ holds. Θ' has the following properties:

1. Since transformation step (3) introduces no spatial variable, Θ' contains the same spatial variables as Θ.

2. For any variable $v_L \in \mathcal{V}$ the four corresponding spatial variables are related as shown in Figure 6.3(b) or 6.3(c).

3. For any literal occurrence $l_{i,j}$ the two corresponding spatial variables together with two spatial variables of the affiliated variable are in a form as shown in Figure 6.3(b) or 6.3(c).

4. Only the relations $\{$PP$\}, \{$PP$^{-1}\}, \{$PO$\}$ and $\{*\}$ occur in Θ'

5. If $x\{$PP$\}y$ holds for a spatial variable x and a spatial variable y, then there is no spatial variable z in Θ' with $z\{$PP$\}x$.

6. If $x\{$PP$^{-1}\}y$ holds for a spatial variable x and a spatial variable y, then there is no spatial variable z in Θ' with $z\{$PP$^{-1}\}x$.

Because of (5) and (6), the spatial variables can be divided into two sets. The "small" set S contains all spatial variables that are proper part of other spatial variables, the "big" set B contains all spatial variables that are converse proper part of other spatial variables. The relation PO only holds between regions of the same set. All other relations are unspecified. For proving that Θ' is consistent, we will give a spatial configuration M' that holds all the specified relations and therefore is a model for Θ'. In M' every spatial variable x_i of the small set S is instantiated by a "small region" S_i', every spatial variable x_i of the big set B is instantiated by a "big region" B_i'.

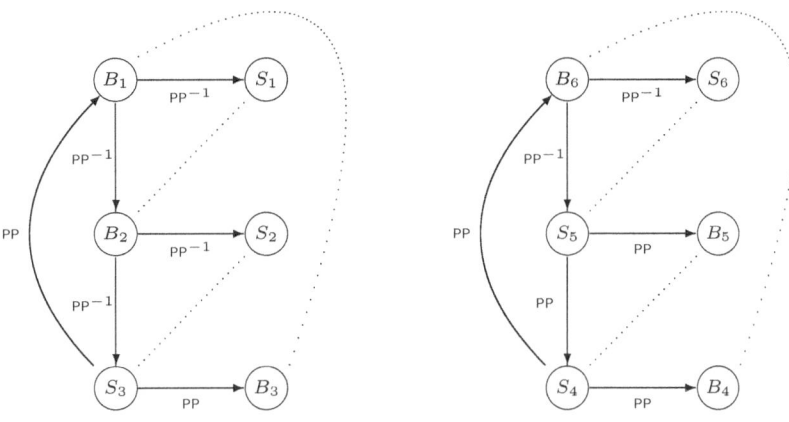

Fig. 6.5. Two different consistent refinements of not-all-equal clauses to base relations

- Every small region $S_i' \in M'$ consists of two parts s' and s_i', where s' is common to all small regions. $s_i'\{\mathsf{DR}\}s_j'$ for all $i \neq j$ and $s_i'\{\mathsf{DR}\}s'$ for all $i = 1\ldots|S|$.
- Every big region $B_i' \in B$ consists of two parts b' and b_i', where b' is common to all big regions. $b_i'\{\mathsf{DR}\}b_j'$ for all $i \neq j$ and $b_i'\{\mathsf{DR}\}b'$ for all $i = 1\ldots|B|$.
- For all $i = 1\ldots|S|$: $s_i'\{\mathsf{PP}\}b'$ and $s'\{\mathsf{PP}\}b'$.

As every small region is proper part of every big region and all small regions as well as all big regions partially overlap, Θ' is consistent. Θ results from Θ' after applying transformation step (3). The spatial configuration M' is no model for Θ, because two relations of every clause do not hold. This can be seen in Figure 6.5 which shows two possible consistent refinements of a clause and the instantiations of the spatial variables with small and big regions. In every clause there is either one small region which is proper part of another small region or a big region which is proper part of another big region. In Figure 6.5 this is the case for $B_1\{\mathsf{PP}^{-1}\}B_2$ and $S_4\{\mathsf{PP}^{-1}\}S_5$. Also there is one small region in every clause that partially overlaps one big region. In Figure 6.5 this is the case for $B_2\{\mathsf{PO}\}S_1$ and $B_5\{\mathsf{PO}\}S_4$.

With a few changes to M', we can construct a spatial configuration M that is a model for Θ. In M every spatial variable x_i of the small set is instantiated by S_i, every spatial variable x_i of the big set is instantiated by B_i. Apart from the following exceptions S_i is equal to S_i' and B_i is equal to B_i':

- For any $i, j = 1\ldots|S|$: If $S_i\{\mathsf{PP}^{-1}\}S_j$ must hold then $S_i := S_i' \cup s_j'$.

– For any $i, j = 1 \ldots |B|$: If $B_i\{\mathsf{PP}^{-1}\}B_j$ must hold then $B_i := B_i' \cup b_j'$.

– For any $i = 1 \ldots |B|, j = 1 \ldots |S|$: If $B_i\{\mathsf{PO}\}S_j$ must hold then $B_i := B_i' \backslash s_j'$.

As exactly two exceptions occur in every clause and the literal occurrences of every clause correspond to different regions, no region will be changed more than once. The regions of M hold all relations of Θ.

The transformation takes time linear in the number of clauses, so $\mathsf{RSAT}(\Theta)$ is NP-hard. ∎

Since RCC-5 is a subset of RCC-8, this result can be easily applied to RCC-8. When $\{\mathsf{PP}\}$ is replaced by $\{\mathsf{TPP}, \mathsf{NTPP}\}$ and the same for their converse, Properties 6.1 and 6.2 hold accordingly, and the same proof can be carried out.

Corollary 6.4. $\mathsf{RSAT}(\text{RCC-8})$ *is NP-hard.*

Since we now know that RSAT is NP-hard, we want to find out whether there are subsets of RCC-8 for which the consistency problem can be decided in polynomial time, and ideally identify the borderline between tractability and intractability. In order to identify this borderline, one has to examine all subsets of RCC-8. We limit ourselves to subsets containing all base relations, because these subsets still allow one to express definite knowledge, if it is available. Additionally, we require the universal relation to be in the subset, so that it is possible to express complete ignorance. This reduces the number of subsets we have to analyze from 2^{256} to 2^{247}. Fortunately, we can reduce the number of subsets further by noting that the computational complexity associated with an arbitrary subset $\mathcal{S}$ is identical to the complexity associated with the closure of this subset under composition, intersection, and converse, denoted by $\widehat{\mathcal{S}}$ – an observation that has also been used in determining a maximal tractable subset of Allen's interval calculus [128, Theorem 14]. Instead of reproving this fact for the RCC-8 relations, we will prove a more general result.

Theorem 6.5. *Let $\mathcal{C}$ be a set of binary relations that is closed under composition, intersection, and converse. Then for any subset $\mathcal{S} \subseteq \mathcal{C}$ that contains the universal relation, the problem $\mathsf{CSPSAT}(\widehat{\mathcal{S}})$ can be polynomially reduced to $\mathsf{CSPSAT}(\mathcal{S})$.*

Proof. Let $\mathcal{T} = \widehat{\mathcal{S}} \setminus \mathcal{S}$. Every element $R \in \mathcal{T}$ can be expressed by successive application of composition, intersection and converse of elements of $\mathcal{S}$. Let n be the maximal number of operations needed for a single element, where for each element the minimal number of operations is considered.

We will show by induction that for any set Θ of constraints over $\widehat{\mathcal{S}}$ we can construct a set Θ' of constraints over $\mathcal{S}$ with $|\Theta'| \leq 2^n \times |\Theta|$ and Θ consistent if and only if Θ' consistent (induction hypothesis). Since n is fixed, the transformation is polynomial.

Base step ($n = 1$): Θ' contains all constraints $(x\,S\,y) \in \Theta$ with $S \in \mathcal{S}$. For any constraint $(x\,R\,y) \in \Theta$ with $R \in \mathcal{T}$ one of the following cases applies:

1. $R = S^{\smile}$ and $S \in \mathcal{S}$. Add $(y\,S\,x)$ to Θ'.

2. $R = S \circ T$ and $S, T \in \mathcal{S}$. Add $(x\,S\,z)$ and $(z\,T\,y)$ to Θ', where z is a fresh variable.

3. $R = S \cap T$ and $S, T \in \mathcal{S}$. Add $(x\,S\,y)$ and $(x\,T\,y)$ to Θ'.

Then Θ' is consistent if and only if Θ is consistent, and $|\Theta'| \leq 2^1 \times |\Theta|$.

Inductive step: Suppose that $n = k+1$ and that the induction hypothesis holds for $n = k$. Let $\mathcal{T}' \subseteq \mathcal{T}$ be the set of relations that can only be composed of relations of $\mathcal{S}$ using $k+1$ operations. Then $\widehat{\mathcal{S}} \setminus \mathcal{T}'$ is the set of relations that can be composed of $\mathcal{S}$ using a maximal number of k operations. So the induction hypothesis is valid for $\widehat{\mathcal{S}} \setminus \mathcal{T}'$. Any relation $R \in \mathcal{T}'$ can be composed of relations of $\widehat{\mathcal{S}} \setminus \mathcal{T}'$ using exactly one operation, so R can be treated as specified in the base step. ∎

Note that Theorem 6.5 holds only if there exists an infinite supply of fresh variables; this is not always the case (e.g., bounded variable problems which are studied in logic and model theory). Another requirement of Theorem 6.5 is the possibility to specify more than one constraint for each pair of variables. Otherwise the identity relation must be contained in $\mathcal{S}$. Since RSAT is a special case of CSPSAT, Theorem 6.5 can also be applied to RSAT.

Corollary 6.6. *Let $\mathcal{S}$ be a subset of* RCC-8.

1. RSAT$(\widehat{\mathcal{S}}) \in$ P *if and only if* RSAT$(\mathcal{S}) \in$ P.

2. RSAT$(\mathcal{S})$ *is* NP-*hard if and only if* RSAT$(\widehat{\mathcal{S}})$ *is* NP-*hard.*

The first statement of Corollary 6.6 can be used to increase the number of elements of tractable subsets of RCC-8 considerably. With the second statement of Corollary 6.6 NP-hardness proofs of RSAT can be used to exclude certain relations from being in any tractable subset of RCC-8. The NP-hardness proof of Theorem 6.3, for example, only contains the relations {PO} and {TPP, TPP^{-1}, NTPP, NTPP^{-1}} when written in RCC-8 relations. So for any subset $\mathcal{S}$ with the two relations contained in its closure $\widehat{\mathcal{S}}$, RSAT$(\mathcal{S})$ is NP-hard.

A further NP-hardness proof of RSAT(RCC-8) can be specified to exclude more relations from being in a tractable subset of RCC-8. This proof uses a polynomial transformation from ONE-IN-THREE-3SAT where $R_t = \{\text{NTPP}\}$ and $R_f = \{\text{TPP}^{-1}\}$. Three more properties, which can be verified using the composition table (Table 4.2), are necessary for specifying the clause and the polarity constraints.

Property 6.7. *Let x, y, z be spatial variables, where $x\{$NTPP, TPP$^{-1}\}y$, $y\{$NTPP, TPP$^{-1}\}z$ and $z\{$NTPP, TPP$^{-1}\}x$ hold. A refinement of these relations to base relations is consistent only when exactly one of the relations is refined to $\{$NTPP$\}$.*

Property 6.8. *Let x, y, z be spatial variables, where $x\{$NTPP, TPP$^{-1}\}z$, $y\{$NTPP, TPP$^{-1}\}z$ and $x\{$PO$\}y$ hold. A refinement of these relations to base relations is consistent only when the relations between (x, z) and between (y, z) are refined to the same base relation.*

Property 6.9. *Let x, y, z be spatial variables, where $x\{$NTPP, TPP$^{-1}\}z$, $z\{$NTPP, TPP$^{-1}\}y$ and $x\{$PO$\}y$ hold. A refinement of these relations to base relations is consistent only when the relations between (x, z) and between (z, y) are refined to different base relation.*

Property 6.7 will be used for the one-in-three condition of the literals in a clause, Properties 6.8 and 6.9 are used to propagate a refinement of a constraint to another constraint.

Lemma 6.10. *Let $\mathcal{S}$ be a subset of RCC-8 containing all base relations. If any of the relations $\{$TPP, NTPP, TPP^{-1}, NTPP$^{-1}\}$, $\{$TPP, TPP$^{-1}\}$, $\{$NTPP, NTPP$^{-1}\}$, $\{$NTPP, TPP$^{-1}\}$ or $\{$TPP, NTPP$^{-1}\}$ is contained in $\widehat{\mathcal{S}}$, then RSAT$(\mathcal{S})$ is NP-hard.*

Proof. NP-hardness of RSAT$(\mathcal{B} \cup \{\{$TPP, NTPP, TPP^{-1}, NTPP$^{-1}\}\})$ can be proved by replacing the RCC-5 relations of Theorem 6.3 with the corresponding RCC-8 relations. NP-hardness of RSAT$(\mathcal{B} \cup \{\{$TPP, TPP$^{-1}\}\})$ and RSAT$(\mathcal{B} \cup \{\{$NTPP, NTPP$^{-1}\}\})$ can be proved by replacing $\{$PP$\}$ with $\{$TPP$\}$ and $\{$PP$^{-1}\}$ with $\{$TPP$^{-1}\}$ or by replacing $\{$PP$\}$ with $\{$NTPP$\}$ and $\{$PP$^{-1}\}$ with $\{$NTPP$^{-1}\}$, respectively. Then Properties 6.1 and 6.2 hold accordingly, so the transformation of Theorem 6.3 can also be applied using $\{$TPP, TPP$^{-1}\}$ or $\{$NTPP, NTPP$^{-1}\}$ instead of $\{$PP, PP$^{-1}\}$. A model M for a set of RCC-8-constraints Θ obtained from this revised transformation can be constructed in the same way as specified in Theorem 6.3.

In order to prove NP-hardness of RSAT$(\mathcal{B} \cup \{\{$NTPP, TPP$^{-1}\}\})$, Properties 6.7, 6.8, and 6.9 are required. Then ONE-IN-THREE-3SAT can be polynomially transformed to RSAT with the same transformation steps as specified in Theorem 6.3. Within these steps $\{$PP$\}$ has to be replaced with $\{$NTPP$\}$ and $\{$PP$^{-1}\}$ with $\{$TPP$^{-1}\}$. The effect of the polarity constraints and the clause constraints can be seen in Figures 6.6 and 6.7. Because of Property 6.7 exactly one literal must be true in any clause.

We will now construct a model M for a set of RCC-8-constraints Θ resulting from the transformation from a satisfiable instance of ONE-IN-THREE-3SAT. As in Theorem 6.3, it can be distinguished between "small" regions S_i and "big" regions B_i where a spatial variable X_i is instantiated with S_i if $X_i\{$NTPP$\}Y_i$ holds and is instantiated with B_i if $X_i\{$TPP$^{-1}\}Y_i$ holds. If X_i is instantiated with S_i then Y_i is instantiated with B_i and *vice versa*. It can

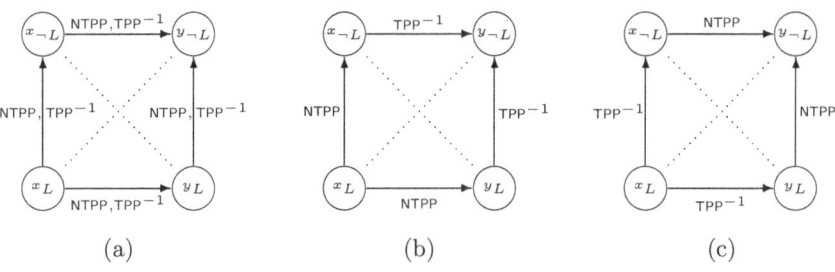

Fig. 6.6. The polarity constraints (a) for the transformation of ONE-IN-THREE-3SAT assure that positive and negative literals of the same variable have opposite assignments: (b) and (c) are the only consistent refinements to base relations

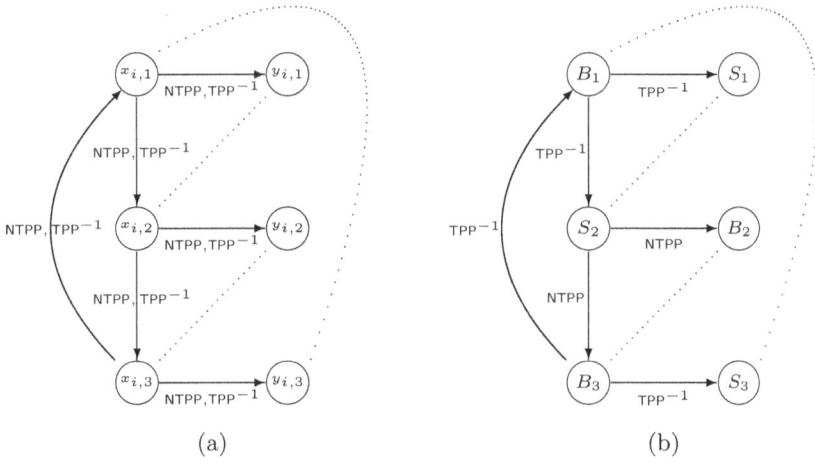

Fig. 6.7. Transformation of a one-in-three clause $c_i = \{l_{i,1}, l_{i,2}, l_{i,3}\}$ to spatial constraints (a) and a consistent refinement of a the clause to base relations (b)

be seen in Figure 6.7(b) that (with two exceptions) all small regions partially overlap each other and are part of all big regions which, again, partially overlap each other. In every clause there are two exceptions, namely, that one big region is tangential proper part of another big region and one small region partially overlaps one big region. In Figure 6.7(b) this is the case for regions B_1 and B_3 and for regions B_1 and S_3. In order to construct all regions B_i and S_i, we need a region B plus the regions S, s_i, $s_{i,j}$, $s'_{i,j}$, and $b_{i,j}$ for all $i, j = 1, ..., n$ (n is the number of spatial variables in Θ) which are all non-tangential proper part of B. All regions in B are disconnected with all other regions in B except for the following relationships: $s_{i,j}\{EC\}b_{j,i}$ for all i, j and $s'_{i,j}\{NTPP\}b_{j,i}$ for all i, j. Let $B'_i := B \setminus \{b_{i,j}|j\}$ and $S'_i := S \cup s_i$. Then all

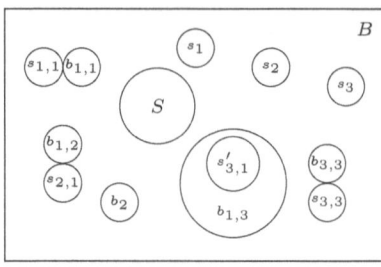

Fig. 6.8. A model for the six regions in Figure 6.7 where $B_1 = B \setminus \{b_{1,1}, b_{1,2}, b_{1,3}, b_{3,3}\}$, $B_2 = B \setminus b_2$, $B_3 = B \setminus b_{3,3}$, $S_1 = S \cup s_1 \cup s_{1,1}$, $S_2 = S \cup s_2 \cup s_{2,1}$, and $S_3 = S \cup s_3 \cup s_{3,3} \cup s'_{3,1}$

regions B'_i partially overlap each other, all regions S'_i partially overlap each other and are non-tangential proper part of every region B'_j. B_i and S_i can be obtained from B'_i and S'_i with some modifications (see Figure 6.8):

- If $S_i\{\mathsf{TPP}\}B_j$ must hold, then $S'_i := S'_i \cup s_{i,j}$.
- If $S_i\{\mathsf{PO}\}B_j$ must hold, then $S'_i := S'_i \cup s'_{i,j}$.
- If $B_i\{\mathsf{TPP}\}B_j$ must hold, then $B'_i := B'_i \setminus \{b_{j,k}|k\}$

After these modifications, all relations required by Θ hold for any B'_i and S'_j, so by setting $B_i := B'_i$ and $S_i := S'_i$ for all i we obtain a model of Θ. Thus, an instance of **ONE-IN-THREE-3SAT** has a solution if and only if the set of RCC-8-constraints obtained by the specified transformation is consistent. The transformation is polynomial, therefore, $\mathsf{RSAT}(\mathcal{B} \cup \{\{\mathsf{NTPP}, \mathsf{TPP}^{-1}\}\})$ is NP-hard.

If $\{\mathsf{TPP}, \mathsf{NTPP}^{-1}\}$ is contained in $\widehat{\mathcal{S}}$, then $\{\mathsf{TPP}^{-1}, \mathsf{NTPP}\}$ is also contained, so $\mathsf{RSAT}(\mathcal{B} \cup \{\{\mathsf{TPP}, \mathsf{NTPP}^{-1}\}\})$ is also NP-hard. With Corollary 6.6 the proof is completed. ∎

By computing the closure of all sets containing the eight base relations together with one additional relation, we obtain the following lemma.

Lemma 6.11. $\mathsf{RSAT}(\mathcal{S})$ *is NP-hard for any subset $\mathcal{S}$ of* RCC-8 *containing all base relations together with one of the 72 relations of the following sets:*

$$\mathcal{N}_1 = \{R \mid \{\mathsf{PO}\} \not\subseteq R \text{ and } (\{\mathsf{TPP}, \mathsf{TPP}^{-1}\} \subseteq R \text{ or } \{\mathsf{NTPP}, \mathsf{NTPP}^{-1}\} \subseteq R)\},$$
$$\mathcal{N}_2 = \{R \mid \{\mathsf{PO}\} \not\subseteq R \text{ and } (\{\mathsf{TPP}, \mathsf{NTPP}^{-1}\} \subseteq R \text{ or } \{\mathsf{TPP}^{-1}, \mathsf{NTPP}\} \subseteq R)\}.$$

Proof. The closure of any of the 72 subsets contains at least one of the five relations of Lemma 6.10. ∎

6.2 Transformation of RSAT to SAT

In the previous section we proved that particular relations cannot be added to the set of RCC-8 base relations without making the consistency problem NP-hard. In order to identify a tractable subset of RCC-8 we have to find out for which set of RCC-8 relations S the consistency problem $\mathsf{RSAT}(S)$ can be reduced to a tractable decision problem. We keep on using propositional satisfiability problems and first transform RSAT to the NP-hard propositional satisfiability problem SAT [31], the problem of deciding whether a propositional formula in conjunctive normal form (CNF) is satisfiable. In the next section we will then determine for which subsets S of RCC-8 $\mathsf{RSAT}(S)$ is reduced to tractable fragments of SAT using the transformation developed in this section. In particular we will use $\mathsf{HORNSAT}$, the tractable problem of deciding satisfiability of propositional Horn formulas, i.e., formulas where each clause contains at most one positive literal.

For transforming RSAT to SAT we transform every instance Θ of RSAT to a propositional formula in CNF that is satisfiable if and only if Θ is consistent. We start with analyzing $m(\Theta)$, the modal encoding of Θ, and then show that whenever $m(\Theta)$ is satisfiable, it has a Kripke model of a specific type. This model is then used to transform $m(\Theta)$ to a classical propositional formula.

6.2.1 Analysis of the Modal Encoding

In this subsection we analyze the modal encoding of RCC-8 and bring it in a form which is suitable for the further transformation to classical propositional logic. Using the modal encoding of RCC-8 given in Section 4.3, a set of RCC-8 constraints Θ can be transformed to a modal formula $m(\Theta)$ as follows, where $Var(\Theta)$ is the set of spatial variables used in Θ:

$$m(\Theta) = \left(\bigwedge_{xRy \in \Theta} m_1(xRy) \right) \wedge \left(\bigwedge_{x \in Var(\Theta)} m_2(x) \right).$$

m_2 consists of the modal formulas that have to be true for every spatial variable $x \in Var(\Theta)$. It results from the regularity constraint (4.27) and the non-emptiness constraint $\neg X$. Instead of (4.27) we use the regularity constraints $\Box(\neg X \to I \neg X) \wedge \Box(X \to \neg I \neg IX)$ which are equivalent to (4.27) in S4.

$$m_2(x) \quad = \quad \Box(\neg X \to I \neg X) \wedge \Box(X \to \neg I \neg IX) \wedge \neg \Box \neg X.$$

The modal encoding m_1 of a spatial constraint xRy is determined by the base relations B contained in R:

$$m_1(xRy) = \bigvee_{B \subseteq R} m_1(xBy).$$

The modal encoding of spatial constraints containing only base relations results directly from Table 4.4:

$$
\begin{aligned}
m_1(x\{DC\}y) &= \Box(\neg(X \land Y)), \\
m_1(x\{EC\}y) &= \Box(\neg(IX \land IY)) \land \neg\Box(\neg(X \land Y)), \\
m_1(x\{PO\}y) &= \neg\Box(\neg(IX \land IY)) \land \neg\Box(X \to Y) \land \neg\Box(Y \to X), \\
m_1(x\{TPP\}y) &= \Box(X \to Y) \land \neg\Box(X \to IY) \land \neg\Box(Y \to X), \\
m_1(x\{TPP^{-1}\}y) &= \Box(Y \to X) \land \neg\Box(Y \to IX) \land \neg\Box(X \to Y), \\
m_1(x\{NTPP\}y) &= \Box(X \to IY) \land \neg\Box(Y \to X), \\
m_1(x\{NTPP^{-1}\}y) &= \Box(Y \to IX) \land \neg\Box(X \to Y), \\
m_1(x\{EQ\}y) &= \Box(X \to Y) \land \Box(Y \to X).
\end{aligned}
$$

As follows from the work by Bennett [12, 13], Θ is consistent if and only if $m(\Theta)$ is satisfiable. It is striking that $m(\Theta)$ is composed only of conjunctions and disjunctions of the model and entailment constraints and the regularity constraints without using any other modal operators to combine them. Thus, a classical propositional formula in CNF can be obtained from $m(\Theta)$ by using the following steps:

1. Consider the model and entailment constraints and the regularity constraints as "propositional atoms".
2. Transform $m(\Theta)$, which is a conjunction and disjunction of these "propositional atoms", into CNF.
3. Transform the disjunctions of "propositional atoms" into propositional formulas in CNF.

In order to make the following application of these steps more readable, we will introduce some abbreviations for the model and entailment constraints and for the regularity constraints.

Definition 6.12 (abbreviations for modal constraints).
Abbreviations for the model constraints:[2]

$$
\begin{array}{rclcrcl}
\delta_{xy} &\equiv& \Box(\neg(X \land Y)) & & \delta i_{xy} &\equiv& \Box(\neg(IX \land IY)) \\
\pi_{xy} &\equiv& \Box(X \to Y) & & \pi i_{xy} &\equiv& \Box(X \to IY) \\
\gamma_{xy} &\equiv& \Box(Y \to X) & & \gamma i_{xy} &\equiv& \Box(Y \to IX).
\end{array}
$$

Abbreviations for the regularity constraints:[3]

$$
CP_x \equiv \Box(\neg X \to I\neg X) \quad\big|\quad RP_x \equiv \Box(X \to \neg I\neg IX)
$$

---

[2] The abbreviations are the first letters of the meaning of the constraints written in Greek symbols. The constraint $\Box(\neg(X \land Y))$ means that X and Y are Disconnected, hence *delta* (δ); $\Box(X \to Y)$ means that X is Part of Y, hence *pi* (π); and $\Box(Y \to X)$ means that X Contains Y, hence *gamma* (γ). The "i" in the three abbreviations δi, πi, and γi indicates that the interior of regions is involved in the constraints.

[3] The abbreviations indicate the meaning of the two properties, namely, the Closedness Property (CP) and the Regularity Property (RP).

As the entailment constraints are negations of the model constraints, they will be abbreviated as negations of the above abbreviations, e.g., $\neg\delta$ is the abbreviation for the entailment constraint $\neg\Box(\neg(X\wedge Y))$. The entailment constraint $\neg\Box\neg X$ (non-emptiness constraint) can be written as a combination of abbreviations, namely, $\neg\delta_{xy}\vee\neg\pi_{xy}$. When it is obvious which atoms (regions) are used, the abbreviations will be written without indices.

Now it is possible to write $m(\Theta)$ using only conjunctions and disjunctions of abbreviations of Definition 6.12. We will call this form of writing $m(\Theta)$ the *abbreviated form* of $m(\Theta)$. The *abbreviated form of a relation* R is the modal encoding $m_1(xRy)$ of R (for some spatial variables x and y) written in abbreviated form. The abbreviated form of the eight base relations is the following:

$$
\begin{aligned}
m_1(x\{\mathsf{DC}\}y) &\equiv \delta \\
m_1(x\{\mathsf{EC}\}y) &\equiv \neg\delta \wedge \delta i \\
m_1(x\{\mathsf{PO}\}y) &\equiv \neg\pi \wedge \neg\gamma \wedge \neg\delta i \\
m_1(x\{\mathsf{TPP}\}y) &\equiv \pi \wedge \neg\gamma \wedge \neg\pi i \\
m_1(x\{\mathsf{TPP}^{-1}\}y) &\equiv \neg\pi \wedge \gamma \wedge \neg\gamma i \\
m_1(x\{\mathsf{NTPP}\}y) &\equiv \neg\gamma \wedge \pi i \\
m_1(x\{\mathsf{NTPP}^{-1}\}y) &\equiv \neg\pi \wedge \gamma i \\
m_1(x\{\mathsf{EQ}\}y) &\equiv \pi \wedge \gamma
\end{aligned}
$$

The abbreviated form of the other relations can be obtained by disjunctively connecting the abbreviated form of the contained base relations. We can now regard the abbreviations as "propositional atoms" and write the abbreviated form of $m(\Theta)$ in conjunctive normal form with respect to these "propositional atoms".

6.2.2 Determining a Particular Kripke Model

In order to transform the modal encoding $m(\Theta)$ of Θ to a classical propositional formula, we must find a finite Kripke frame by which $m(\Theta)$ can be modeled if it is satisfiable. With respect to this Kripke frame, $m(\Theta)$ will then be transformed to a classical propositional formula. Since the transformation must be polynomial in n, the number of spatial variables of Θ, the Kripke frame, i.e., the number of worlds of the frame must be polynomial in n.

Before identifying a particular Kripke frame, we will first have a look at the conditions that must be satisfied if $m(\Theta)$ is satisfiable. $m(\Theta)$ is satisfiable if it is true in a world w of a Kripke model $\mathcal{M} = \langle W, \{R_\Box = W \times W, R_{\mathbf{I}} \subseteq W \times W\}, \nu\rangle$, where W is a set of worlds, $R_\Box$ the accessibility relation of the $\Box$-operator, $R_{\mathbf{I}}$ the accessibility relation of the **I**-operator, and ν a valuation that assigns a truth value to every atom in every world. The truth conditions for $\mathcal{M}, w \Vdash m(\Theta)$ can be specified as a combination of truth conditions of sub-formulas according to the form of $m(\Theta)$:[4]

$$\mathcal{M}, w \Vdash m(\Theta) \quad \text{iff} \quad \mathcal{M}, w \Vdash \bigwedge_{xRy\in\Theta} m_1(xRy) \wedge \mathcal{M}, w \Vdash \bigwedge_{x\in Var(\Theta)} m_2(x)$$

$$\text{iff} \quad \bigwedge_{xRy\in\Theta} \mathcal{M}, w \Vdash m_1(xRy) \wedge \bigwedge_{x\in Var(\Theta)} \mathcal{M}, w \Vdash m_2(x).$$

Since $m(\Theta)$ is composed only of conjunctions of model and entailment constraints and regularity constraints as specified in Section 4.3, we can carry on expressing the truth conditions of $m(\Theta)$ by combining the truth conditions of sub-formulas until we have only combinations of formulas of the type $\mathcal{M}, w \Vdash \Box\varphi$ or $\mathcal{M}, w \Vdash \neg\Box\varphi$. These sub-formulas correspond to the different modal and entailment constraints and the regularity constraints and form 15 structurally different formulas. We express them in terms of the abbreviations as given in Definition 6.12. The truth conditions of these formulas can be obtained by combining the truth values of the single atoms. We start with specifying the truth conditions of the model constraints.[5]

$$\mathcal{M}, w \Vdash \delta \quad \text{iff} \quad \forall u. (\mathcal{M}, u \not\Vdash X \vee \mathcal{M}, u \not\Vdash Y) \tag{6.1}$$

$$\mathcal{M}, w \Vdash \pi \quad \text{iff} \quad \forall u. (\mathcal{M}, u \not\Vdash X \vee \mathcal{M}, u \Vdash Y) \tag{6.2}$$

$$\mathcal{M}, w \Vdash \gamma \quad \text{iff} \quad \forall u. (\mathcal{M}, u \Vdash X \vee \mathcal{M}, u \not\Vdash Y) \tag{6.3}$$

$$\mathcal{M}, w \Vdash \delta i \quad \text{iff} \quad \forall u.(\mathcal{M}, u \not\Vdash IX \vee \mathcal{M}, u \not\Vdash IY)$$

$$\text{iff} \quad \forall u.\exists v : uR_Iv.(\mathcal{M}, v \not\Vdash X \vee \mathcal{M}, v \not\Vdash Y) \tag{6.4}$$

$$\mathcal{M}, w \Vdash \pi i \quad \text{iff} \quad \forall u.(\mathcal{M}, u \not\Vdash X \vee \mathcal{M}, u \Vdash IY)$$

$$\text{iff} \quad \forall u.\forall v : uR_Iv.(\mathcal{M}, u \not\Vdash X \vee \mathcal{M}, v \Vdash Y) \tag{6.5}$$

$$\mathcal{M}, w \Vdash \gamma i \quad \text{iff} \quad \forall u.(\mathcal{M}, u \Vdash IX \vee \mathcal{M}, u \not\Vdash Y)$$

$$\text{iff} \quad \forall u.\forall v : uR_Iv.(\mathcal{M}, v \Vdash X \vee \mathcal{M}, u \not\Vdash Y) \tag{6.6}$$

Now the truth conditions of the entailment constraints.

$$\mathcal{M}, w \Vdash \neg\delta \quad \text{iff} \quad \exists u. (\mathcal{M}, u \Vdash X \wedge \mathcal{M}, u \Vdash Y) \tag{6.7}$$

$$\mathcal{M}, w \Vdash \neg\pi \quad \text{iff} \quad \exists u. (\mathcal{M}, u \Vdash X \wedge \mathcal{M}, u \not\Vdash Y) \tag{6.8}$$

$$\mathcal{M}, w \Vdash \neg\gamma \quad \text{iff} \quad \exists u. (\mathcal{M}, u \not\Vdash X \wedge \mathcal{M}, u \Vdash Y) \tag{6.9}$$

$$\mathcal{M}, w \Vdash \neg\delta \vee \neg\gamma \quad \text{iff} \quad \exists u. (\mathcal{M}, u \Vdash X) \tag{6.10}$$

$$\mathcal{M}, w \Vdash \neg\delta i \quad \text{iff} \quad \exists u.(\mathcal{M}, u \Vdash IX \wedge \mathcal{M}, u \Vdash IY)$$

$$\text{iff} \quad \exists u.\forall v : uR_Iv.(\mathcal{M}, v \Vdash X \wedge \mathcal{M}, v \Vdash Y) \tag{6.11}$$

$$\mathcal{M}, w \Vdash \neg\pi i \quad \text{iff} \quad \exists u.(\mathcal{M}, u \Vdash X \wedge \mathcal{M}, u \not\Vdash IY)$$

---

[4] Note that the logical operators used to combine expressions of the form $\mathcal{M}, w \Vdash \varphi$ are not the usual modal operators, but operators in a meta language.

[5] Since $\Box$ is a strong S5-operator and every world is accessible with $R_\Box$ from any other world, the condition *for all u with $wR_\Box u$* can be replaced with $\forall u$.

$$\text{iff}\quad \exists u.\exists v : uR_\mathbf{I}v.(\mathcal{M}, u \Vdash X \wedge \mathcal{M}, v \nVdash Y) \quad (6.12)$$

$$\mathcal{M}, w \Vdash \neg\gamma i \quad \text{iff}\quad \exists u.(\mathcal{M}, u \nVdash \mathbf{I}X \wedge \mathcal{M}, u \Vdash Y)$$

$$\text{iff}\quad \exists u.\exists v : uR_\mathbf{I}v.(\mathcal{M}, v \nVdash X \wedge \mathcal{M}, u \Vdash Y) \quad (6.13)$$

Finally, the truth conditions of the regularity constraints.

$$\mathcal{M}, w \Vdash \text{CP} \quad \text{iff}\quad \forall u.(\mathcal{M}, u \Vdash X \vee \mathcal{M}, u \Vdash \mathbf{I}\neg X)$$

$$\text{iff}\quad \forall u.\forall v : uR_\mathbf{I}v.(\mathcal{M}, u \Vdash X \vee \mathcal{M}, v \nVdash X) \quad (6.14)$$

$$\mathcal{M}, w \Vdash \text{RP} \quad \text{iff}\quad \forall u.(\mathcal{M}, u \nVdash X \vee \mathcal{M}, u \nVdash \mathbf{I}\neg \mathbf{I}X)$$

$$\text{iff}\quad \forall u.\exists s : uR_\mathbf{I}s.\forall t : sR_\mathbf{I}t.(\mathcal{M}, u \nVdash X \vee \mathcal{M}, t \Vdash X) \quad (6.15)$$

We call this form of writing $m(\Theta)$ the *explicit form* of $m(\Theta)$.

We will construct a polynomial Kripke frame $\mathcal{F}$ on which a model $\mathcal{M}$ for $m(\Theta)$ can be based if $m(\Theta)$ is satisfiable at all. For this we will successively add only those worlds to the frame that are explicitly required by the formulas (6.1) – (6.15). Amongst them are formulas of the type $\forall w.\varphi$ and formulas of the type $\exists w.\varphi$. Only formulas of the type $\exists w.\varphi$ explicitly require a world with the specified properties φ. If a world with these properties is not present, it has to be introduced. Formulas of the type $\forall w.\varphi$ affect the properties of worlds already introduced, but do not require fresh worlds. These properties must be enforced by the model and not by the frame. So worlds have to be introduced only for existential quantifiers. Since there are also formulas of the type $\forall w.\exists v.\phi$ a fresh world must not be introduced for every occurring existential quantifier. In the following we will analyze which existential quantifiers introduce fresh worlds and for which existential quantifiers there already exist worlds with the specified properties. For this we will classify worlds according to the accessibility relation $R_\mathbf{I}$ holding between them.

Definition 6.13 (level of worlds).
Let $u, v, w \in W$ be worlds of the Kripke frame $\mathcal{F}$.

– *u is a world of* level 0 *if $vR_\mathbf{I}u$ only holds for $v = u$.*

– *u is a world of* level $l+1$ *if $vR_\mathbf{I}u$ holds for a world v of level l and if there is no world $w \neq u$ of a level higher than l with $wR_\mathbf{I}u$.*

– *If w is a world of level 0 and v is an $R_\mathbf{I}$-successor of w, then w is called* the *introductory world of v.*

– *$W_i \subseteq W$ is the set of worlds of level i.*

Every occurrence of one of the formulas (6.7) to (6.13), which correspond to the six different entailment constraints, shall introduce a fresh world of level 0. Formulas (6.12) and (6.13) introduce additional worlds of level 1. Every model shall be designed according to these principles. Instead of analyzing which world requires a fresh successor by the $\forall w \exists v$ condition of formulas (6.5) and (6.13), we will in the following define a Kripke frame of a particular structure and prove that whenever $m(\Theta)$ is consistent there is a model based on that frame.

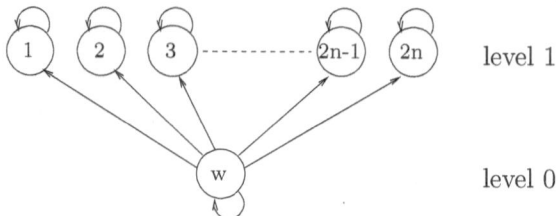

Fig. 6.9. A world w of level 0 together with its $2n$ $R_\mathbf{I}$-successors of an RCC-8-frame of level 1. Worlds are drawn as circles, the arrows indicate the accessibility of worlds with the relation $R_\mathbf{I}$

Definition 6.14 (RCC-8-frame, RCC-8-model).
Let Θ be a set of RCC-8-constraints and $n = |Var(\Theta)|$ be the number of regions in Θ. An RCC-8-frame $\mathcal{F}_l = \langle W, \{R_\square, R_\mathbf{I}\}\rangle$ of level l has the following properties (see Figure 6.9):

1. *W contains only worlds up to level l, i.e., $W = \bigcup_{i=0}^{l} W_i$*

2. *For every world $u \in W_i$ there are exactly $2n$ worlds $v \in W_{i+1}$ with $uR_\mathbf{I}v$.*

3. *For every world $v \in W_k$ there is exactly one world $u \in W_i$ with $uR_\mathbf{I}v$ for every level $0 \le i \le k$ (if $i = k$ then $u = v$).*

4. *For all worlds $u, v, w \in W: wR_\square v, wR_\mathbf{I}w$, and $wR_\mathbf{I}v, vR_\mathbf{I}u$ implies $wR_\mathbf{I}u$.*

An RCC-8-model of level l is based on an RCC-8-frame of level l. In a poly-nomial RCC-8-frame/-model of level l the number of worlds is polynomially bounded by n.

Note that item (4) guarantees that every RCC-8-frame of level l is an S4-frame. We will now show that RCC-8-frames of level 1 are sufficient for mo-deling every satisfiable formula $m(\Theta)$.

Lemma 6.15. *If $m(\Theta)$ is satisfiable, there is a polynomial RCC-8-model $\mathcal{M}$ of level 1 with $\mathcal{M}, w \Vdash m(\Theta)$.*

Proof The abbreviated form of every relation contains at most 6 different negated abbreviations.[6] Since they correspond to entailment constraints, each of these abbreviations introduces a new world of level 0. As there are n regions and therefore n^2 relations, there are at most $6n^2$ worlds of level 0. For each world w and each atom X there might be a sub-formula of $m(\Theta)$ that forces the existence of an $R_\mathbf{I}$-successor of w that forces X or that forces $\neg\mathsf{X}$. Thus, since there are n different atoms X, a maximal number of $2n$ different $R_\mathbf{I}$-successors is sufficient for each world. Therefore, if $m(\Theta)$ is satisfiable, it can be modeled by an RCC-8-model of some level l.

---

[6] This number results from a straightforwardly computed abbreviated form of every relation. It might be decreased by optimizing the abbreviated form with respect to the number of negated abbreviations.

Suppose that $\mathcal{M}' = \langle W', \{R'_\mathbf{I}, R'_\square\}, \nu' \rangle$ is such an RCC-8-model that models $m(\Theta)$ and suppose there is no polynomial RCC-8-model of level 1 that models $m(\Theta)$. We will prove by contradiction that a polynomial RCC-8-model of level 1 exists if $m(\Theta)$ is satisfiable. For this we will construct a polynomial RCC-8-model $\mathcal{M} = \langle W, R, \nu \rangle$ of level 1 using $\mathcal{M}'$. Let $W = W'_0 \cup W'_1$, $R_\mathbf{I} = \{(u, v) \in R'_\mathbf{I} | u, v \in W\}$, $R_\square = W \times W$, and $\nu(w, a) = \nu'(w, a)$ for all $w \in W$ and all propositional atoms a. We now have to find out which of the formulas (6.1) to (6.15) hold for $\mathcal{M}'$ but not for $\mathcal{M}$. Trivially, if the formulas (6.7) to (6.13), corresponding to the entailment constraints, hold for $\mathcal{M}'$ they also hold for $\mathcal{M}$. The same for the formulas (6.1) to (6.3). As the only $R_\mathbf{I}$-successor of a world of level 1 is the world itself, formulas (6.5), (6.6) and formulas (6.14) and (6.15) also hold for $\mathcal{M}$ if they hold for $\mathcal{M}'$. So only formula (6.4) remains to be checked.

Suppose that formula (6.4) holds for two atoms X and Y and suppose there is a world $v \in W'_1$ with $\nu'(v, \mathsf{X}) = true$ and $\nu'(v, \mathsf{Y}) = true$ and an $R_\mathbf{I}$-successor of v for which either X or Y is $false$. Then (6.4) is satisfied in $\mathcal{M}'$ but not in $\mathcal{M}$. In this case $\mathcal{M}$ is not a model for $m(\Theta)$, but it would be a model if the truth value of either $\nu(v, \mathsf{X})$ or $\nu(v, \mathsf{Y})$ can be set to $false$ without contradicting any other formula. Therefore it is necessary to find out how an atom can be forced to be $true$ in a world of level 1. Formulas (6.2) and (6.3) force it only when another atom is already forced to be $true$ in the same world. If the atom is forced by one of the formulas (6.5), (6.11), (6.15), or axiom schemata (4.24) to be $true$ in a world then it is also forced to be $true$ in all $R_\mathbf{I}$-successors of this world. So, if both X and Y are forced to be $true$ in v, they must both be $true$ in any $R_\mathbf{I}$-successor of v and $\mathcal{M}'$ cannot be a model for $m(\Theta)$, so one of them, say X, is not forced and $\nu(v, \mathsf{X})$ can be set to $false$. We have constructed a polynomial RCC-8-model of level 1 which contradicts our initial assumption that there is no such model. ∎

In the following we will use the term RCC-8-frame/-model to refer to an RCC-8-frame/-model of level 1.

6.2.3 Transformation to a Classical Propositional Formula

We proved that whenever $m(\Theta)$ is satisfiable, a model for $m(\Theta)$ can be based on a polynomial RCC-8-frame that contains as many worlds of level 0 as entailment constraints are contained in $m(\Theta)$. We will now transform the explicit form of $m(\Theta)$ to a classical propositional formula $p(m(\Theta))$ in CNF such that $p(m(\Theta))$ is satisfiable if and only if $m(\Theta)$ is satisfiable in a polynomial RCC-8-frame $\mathcal{F} = \langle W, \{R_\square, R_\mathbf{I}\} \rangle$. For this we will introduce a propositional atom for every world $w \in W$ and every spatial variable $x \in Var(\Theta)$ such that the atom is true if and only if $\nu(w, \mathsf{X}) = true$. In order to preserve the structure of the RCC-8-frame in propositional logic, two functions have to be defined:

Definition 6.16 (functions f, g).
Let $\mathcal{F} = \langle W, \{R_\square, R_I\} \rangle$ be an RCC-8-frame.

- *$f : W \to W_0$ determines the introductory world of every world,*

- *$g : W \to \emptyset \cup \{1, \dots, 2n\}$ provides all worlds of the same introductory world with a specific order, i.e., if the worlds u and v are distinct, have the same level and $f(u) = f(v)$, then $g(u) \neq g(v)$. $g(u) = \emptyset$, if and only if u has level 0.*

For every world $w \in W$ and every spatial variable $x \in Var(\Theta)$ we introduce the propositional atom $\mathsf{X}_{f(w)}^{g(w)}$. Before transforming $m(\Theta)$ to a classical propositional formula $p(m(\Theta))$ in CNF, we will first transform $m(\Theta)$ in a straightforward manner to a classical propositional formula $q(m(\Theta))$.

Definition 6.17 (transformation q).
The explicit form of $m(\Theta)$ is transformed to a classical propositional formula $q(m(\Theta))$ as follows:

- *$\mathcal{M}, w \Vdash \mathsf{X}$ is transformed to $\mathsf{X}_{f(w)}^{g(w)}$.*

- *$\mathcal{M}, w \nVdash \mathsf{X}$ is transformed to $\neg \mathsf{X}_{f(w)}^{g(w)}$.*

- *The meta operators $\wedge$ and $\vee$ (see Section 6.2.2) are transformed to the propositional operators $\wedge$ and $\vee$, respectively.*

Since all worlds of the RCC-8-frame $\mathcal{F}$ are known, quantifiers can be transformed to propositional operators:

- *A universal quantification of particular worlds results in a conjunction over these worlds.*

- *An existential quantification of a world of level 0 results in a disjunction over all worlds of level 0.*

- *An existential quantification of an R_I–successor of world w results in a disjunction over the $2n$ R_I–successor of w.*

With the specified transformation $\delta, \neg\delta$ and RP, for example, are transformed to the following formulas:

$$q(\delta) \;=\; \bigwedge_{w \in W_0} (\neg\mathsf{X}_w \vee \neg\mathsf{Y}_w) \wedge \bigwedge_{w \in W_0} \bigwedge_{i=1}^{2n} (\neg\mathsf{X}_w^i \vee \neg\mathsf{Y}_w^i)$$

$$q(\neg\delta) \;=\; \bigvee_{w \in W_0} (\mathsf{X}_w \wedge \mathsf{Y}_w),$$

$$q(\mathrm{RP}) \;=\; \bigwedge_{w \in W_0} \bigvee_{i=1}^{2n} (\neg\mathsf{X}_w \vee \mathsf{X}_w^i).$$

Lemma 6.18. *$q(m(\Theta))$ is satisfiable if and only if $m(\Theta)$ is satisfiable.*

Proof. It follows from Theorem 6.15 that if $m(\Theta)$ is satisfiable, there is an RCC-8-model $\mathcal{M}$ based on a particular RCC-8-frame $\mathcal{F}$. Since the RCC-8-frame $\mathcal{F}$ is known, $\mathcal{M}$ is determined by its valuation ν which only depends on the model and entailment constraints and on the regularity constraints contained in $m(\Theta)$. The explicit form of $m(\Theta)$ contains all conditions of the valuation ν of $\mathcal{M}$ explicitly, i.e., only conditions of the form $\mathcal{M}, w \Vdash X$ or $\mathcal{M}, w \nVdash X$ are contained, which correspond to $\nu(X, w) = true$ or $\nu(X, w) = false$, respectively. As there is a one-to-one correspondence between $\nu(X, w)$ and the propositional atom $X_{f(w)}^{g(w)}$ of $q(m(\Theta))$ for every world w and every region X, it follows immediately from the transformation specified in Proposition 6.17 that $q(m(\Theta))$ is satisfiable if and only if $m(\Theta)$ is satisfiable, where $X_{f(w)}^{g(w)} = true$ if and only if $\nu(X, w) = true$. ∎

It is obvious that because of the transformation of existential quantifiers to disjunctions, $q(m(\Theta))$ is not in CNF. Simply transforming $q(m(\Theta))$ to CNF is, however, not favorable for the further reduction to HORNSAT which is done in the next section. Transforming, for example, $q(\neg\delta)$ to CNF results in non-Horn formulas. Also $q(RP)$ is not a Horn formula because of the disjunction obtained from transforming the existential quantifier. Therefore we have to treat the existential quantifiers differently in order to obtain $p(m(\Theta))$ in CNF.

As stated before, the entailment constraints, i.e., the negated abbreviations, introduce fresh worlds of level 0, so instead of a disjunction over all worlds of level 0, a fresh world of W of level 0 together with it's $2n$ R_I-successors will be explicitly considered in $p(m(\Theta))$. Then $\neg\delta$, for example, is transformed to the Horn formula $p(\neg\delta) = X_w \wedge Y_w$ *(for a fresh world w of level 0)*.

Handling existential quantifiers of worlds of level 1 is more difficult. We exploit the fact that for any statement of the form $\exists u.wR_I u : \mathcal{M}, u \Vdash X$ or $\exists u.wR_I u : \mathcal{M}, u \nVdash X$, for all n different spatial variables $X \in Var(\Theta)$, at most $2n$ R_I-successors are necessary to assure that there exists a world that makes this statement true. Since in the RCC-8-model every world of level 1 has $2n$ R_I-successors, we will reserve one of these $2n$ R_I-successors for every statement of this kind. Therefore we need a function $h : Var(\Theta) \cup \overline{Var(\Theta)} \rightarrow \{1, \ldots, 2n\}$ that associates every spatial variable and the negation of every spatial variable with one of the $2n$ R_I-successors. Using this function, the following properties must hold for all R_I-successor:

– If there exists an R_I-successor u of w with $\mathcal{M}, u \Vdash X$, then $X_w^{h(x)} = true$.
– If there exists an R_I-successor u of w with $\mathcal{M}, u \nVdash X$, then $X_w^{h(\neg x)} = false$.
– If the $h(X)$-successor of a world w holds $\neg X$ then every R_I-successor of w holds $\neg X$.
– If the $h(\neg X)$-successor of w holds X then every R_I-successor of w holds X.

To ensure these properties, additional formulas have to be added to $p(m(\Theta))$ for every spatial variable x:

$$\bigwedge_{w \in W_0} \bigwedge_{i=1}^{2n} (X_w^{h(x)} \vee \neg X_w^i) \qquad (6.16)$$

$$\bigwedge_{w \in W_0} \bigwedge_{i=1}^{2n} (\neg X_w^{h(\neg x)} \vee X_w^i) \qquad (6.17)$$

Formula (6.16) ensures the first and third property, formula (6.17) the second and forth property. With these modifications to $q(m(\Theta))$ we obtain a propositional formula $p(m(\Theta))$ in conjunctive normal form. The transformation is given in the following proposition.

Proposition 6.19. $m(\Theta)$ *can be transformed to a propositional formula* $p(m(\Theta))$ *in CNF by transforming all model and entailment constraints and the regularity constraints contained in $m(\Theta)$ in the following way:*

$$p(\delta) = \bigwedge_{w \in W_0} (\neg X_w \vee \neg Y_w) \wedge \bigwedge_{w \in W_0} \bigwedge_{i=1}^{2n} (\neg X_w^i \vee \neg Y_w^i)$$

$$p(\neg \delta) = X_w \wedge Y_w \text{ (for a fresh world } w \text{ of level 0)}$$

$$p(\pi) = \bigwedge_{w \in W_0} (\neg X_w \vee Y_w) \wedge \bigwedge_{w \in W_0} \bigwedge_{i=1}^{2n} (\neg X_w^i \vee Y_w^i)$$

$$p(\neg \pi) = X_w \wedge \neg Y_w \text{ (for a fresh world } w \text{ of level 0)}$$

$$p(\gamma) = \bigwedge_{w \in W_0} (X_w \vee \neg Y_w) \wedge \bigwedge_{w \in W_0} \bigwedge_{i=1}^{2n} (X_w^i \vee \neg Y_w^i)$$

$$p(\neg \gamma) = \neg X_w \wedge Y_w \text{ (for a fresh world } w \text{ of level 0)}$$

$$p(\delta i) = \bigwedge_{w \in W_0} \bigwedge_{i=1}^{?n} (\neg X_w^i \vee \neg Y_w^i)$$

$$p(\neg \delta i) = \bigwedge_{i=1}^{2n} (X_w^i \wedge Y_w^i) \text{ (for a fresh world } w \text{ of level 0)}$$

$$p(\pi i) = \bigwedge_{w \in W_0} \bigwedge_{i=1}^{2n} (\neg X_w \vee Y_w^i)$$

$$p(\neg \pi i) = X_w \wedge \neg Y_w^{h(\neg y)} \text{ (for a fresh world } w \text{ of level 0)}$$

$$p(\gamma i) \quad = \quad \bigwedge_{w \in W_0} \bigwedge_{i=1}^{2n} (X_w^i \vee \neg Y_w)$$

$$p(\neg \gamma i) \quad = \quad \neg X_w^{h(\neg x)} \wedge Y_w \text{ (for a fresh world } w \text{ of level 0)}$$

$$p(\mathsf{CP}) \quad = \quad \bigwedge_{w \in W_0} \bigwedge_{i=1}^{2n} (X_w \vee \neg X_w^i)$$

$$p(\mathsf{RP}) \quad = \quad \bigwedge_{w \in W_0} (\neg X_w \vee X_w^{h(x)})$$

$$p(\neg \pi \vee \neg \gamma) \quad = \quad X_w \text{ (for a fresh world } w \text{ of level 0)}$$

Disjunctions of these constraints are transformed into CNF by transforming the disjunction of the corresponding propositional formulas into CNF. This increases the size of the resulting propositional formulas only polynomially.

Lemma 6.20. *$p(m(\Theta))$ is satisfiable if and only if $q(m(\Theta))$ is satisfiable.*

Proof. Every entailment constraint introduces a fresh world of level 0 which explicitly fulfills the requirements of the entailment constraint. Thus, whenever one world fulfills the requirements of an entailment constraint it is fulfilled by the world explicitly introduced by the entailment constraint and *vice versa.* If each of the $2n$ R_I-successors of a world of level 0 is associated to a different spatial variable or the negation of a spatial variable and the formulas (6.16) and (6.17) are added to $p(m(\Theta))$, then $p(\neg \pi i)$, $p(\neg \gamma i)$, and $p(\mathsf{RP})$ are satisfiable if and only if $q(\neg \pi i)$, $q(\neg \gamma i)$, and $q(\mathsf{RP})$ are satisfiable, respectively. For the model constraints and for the constraint CP, p is equal to q. ∎

Theorem 6.21. RSAT(RCC-8) *can be polynomially transformed to* SAT

Proof. From Lemma 6.18 and Lemma 6.20 it follows that $m(\Theta)$ can be transformed to a propositional formula in conjunctive normal form $p(m(\Theta))$ such that $m(\Theta)$ is satisfiable if and only if $p(m(\Theta))$ is satisfiable. Since every world together with every region corresponds to a literal, which are at most $12n^4$ many, and the number of clauses of $p(m(\Theta))$ is polynomial in the number of worlds of the RCC-8-model of $m(\Theta)$ and the size of $m(\Theta)$, the transformation is polynomial. ∎

As SAT is an NP-complete problem, it follows that RSAT is in NP. Together with Corollary 6.4 this results in the following theorem.

Theorem 6.22. RSAT(RCC-8) *is* NP-*complete.*

6.3 Tractable Subsets of RCC-8

In this section we analyze which relations are transformed to propositional Horn formulas using the transformation of RSAT to SAT as specified in the previous section. Since the propositional satisfiability problem for Horn formulas (HORNSAT) is tractable [62], the set of these relations forms a tractable subset of RCC-8. We will then prove that the set of relations identified in this way is maximal with respect to tractability, i.e., no other relation can be added to the set without losing tractability.

6.3.1 Identifying a Large Tractable Subset of RCC-8

In order to identify the relations transformable to Horn formulas, we study the abbreviated form of every relation which consists of a conjunction of disjunctions of abbreviations, i.e., the abbreviated form of relations can be regarded as a "propositional formula" in CNF where the abbreviations are the "propositional atoms". We have to find the relations whose abbreviated form consists only of "clauses" which are transformable to Horn formulas.

Proposition 6.23. *Applying the transformation p to the model and entailment constraints and to the regularity constraints as specified in Proposition 6.19 leads to Horn formulas. Formulas (6.16) and (6.17) are also Horn formulas.*

With these formulas further Horn formulas can be specified.

Lemma 6.24. *The following disjunctions of abbreviations are transformed to propositional Horn formulas:*

$$\delta \vee C \quad \text{with} \quad C \in \{\pi, \neg\pi, \gamma, \neg\gamma, \delta i, \neg\delta i, \pi i, \neg\pi i, \gamma i, \neg\gamma i\},$$
$$\delta i \vee C \quad \text{with} \quad C \in \{\neg\delta, \pi, \neg\pi, \gamma, \neg\gamma, \pi i, \neg\pi i, \gamma i, \neg\gamma i\}.$$

Proof. The propositional formulas resulting from the model constraints δ and δi are indefinite Horn formulas, i.e., clauses that do not contain any positive literals. Because the propositional formulas resulting from the other constraints are Horn formulas, and the disjunction of an indefinite Horn formula with another Horn formula is again a Horn formula, the lemma holds. ∎

In our framework some disjunctions of model and entailment constraints are tautologies. These disjunctions of abbreviations can be eliminated from the abbreviated form.

Lemma 6.25. *The following modal formulas written in abbreviated form are tautologies in S4 in the presence of the non-emptiness constraints and the constraints for regular closed regions:*

$$\neg\delta \vee \neg\pi, \ \neg\delta \vee \neg\gamma, \ \neg\delta \vee \delta i, \ \pi \vee \neg\pi i, \ \gamma \vee \neg\gamma i,$$
$$\neg\pi \vee \neg\delta i, \ \neg\gamma \vee \neg\delta i, \ \neg\gamma \vee \neg\pi i \vee \gamma i, \ \neg\pi \vee \pi i \vee \neg\gamma i.$$

Proof. We prove that the above given modal formulas are tautologies by showing that the negation of each of these formulas results in a contradiction. This can be done by using, e.g., tableau based proof procedures for modal logic [50].

$\delta \wedge \pi \equiv \Box(\neg(X \wedge Y)) \wedge \Box(X \rightarrow Y)$:
   contradicts the non-emptiness constraint $\neg\Box\neg X$.

$\delta \wedge \gamma \equiv \Box(\neg(X \wedge Y)) \wedge \Box(Y \rightarrow X)$: analogous.

$\delta \wedge \neg\delta i \equiv \Box(\neg(X \wedge Y)) \wedge \neg\Box(\neg(IX \wedge IY))$:
   results in a contradiction when combined with the S4 axiom schemata $IX \rightarrow X$.

$\neg\pi \wedge \pi i \equiv \neg\Box(X \rightarrow Y) \wedge \Box(X \rightarrow IY)$:
   results in a contradiction when combined with the closedness constraint $\Box(\neg X \rightarrow \neg IX)$.

$\neg\gamma \wedge \gamma i \equiv \neg\Box(Y \rightarrow X) \wedge \Box(Y \rightarrow IX)$: analogous.

$\pi \wedge \delta i \equiv \Box(X \rightarrow Y) \wedge \Box(\neg(IX \wedge IY))$:
   the regularity constraint $\Box(X \rightarrow \neg I\neg IX)$ and the non-emptiness constraint $\neg\Box\neg X$ enforce that there is a world u with $\mathcal{M}, u \Vdash IX$. Because of the S4 axiom schemata $IX \rightarrow X$, $\mathcal{M}, u \Vdash X$ also holds. π entails $\mathcal{M}, u \Vdash Y$ and δi entails $\mathcal{M}, u \nVdash IY$. So there is an R_I-successor v of u with $\mathcal{M}, v \nVdash Y$. Because $\mathcal{M}, u \Vdash IX$ holds, $\mathcal{M}, v \Vdash X$ also holds. π entails $\mathcal{M}, v \Vdash Y$, which results in a contradiction.

$\gamma \wedge \delta i \equiv \Box(Y \rightarrow X) \wedge \Box(\neg(IX \wedge IY))$: analogous.

$\gamma \wedge \pi i \wedge \neg\gamma i \equiv \Box(Y \rightarrow X) \wedge \Box(X \rightarrow IY) \wedge \neg\Box(Y \rightarrow IX)$:
   because of $\neg\gamma i$ there is a world u with $\mathcal{M}, u \nVdash IX$ and $\mathcal{M}, u \Vdash Y$. Because of γ, $\mathcal{M}, u \Vdash X$ holds and because of πi, $\mathcal{M}, u \Vdash IY$ holds. Since $\mathcal{M}, u \nVdash IX$ holds, there is an R_I-successor v of u with $\mathcal{M}, v \nVdash X$. So γ results in $\mathcal{M}, v \nVdash Y$ and $\mathcal{M}, u \Vdash IY$ results in $\mathcal{M}, v \Vdash Y$, which is a contradiction.

$\pi \wedge \gamma i \wedge \neg\pi i \equiv \Box(X \rightarrow Y) \wedge \Box(Y \rightarrow IX) \wedge \neg\Box(X \rightarrow IY)$: analogous. ∎

All relations with an abbreviated form using only abbreviations or disjunctions of abbreviations transformable to Horn formulas can be transformed to Horn formulas. In this way 64 relations can be transformed to Horn formulas. They are listed in Appendix A.1 together with their abbreviated form. We call the subset of RCC-8 containing these relations $\mathcal{H}_8$.

Theorem 6.26. RSAT($\mathcal{H}_8$) *can be polynomially reduced to* HORNSAT

Proof. Every constraint using a relation of $\mathcal{H}_8$ is transformable to a Horn formula. So every set Θ of $\mathcal{H}_8$-constraints can be written as a conjunction of the Horn formulas of their elements, which is also a Horn formula. ∎

Thus, $\mathsf{RSAT}(\mathcal{H}_8) \in \mathsf{P}$ and because of Corollary 6.6 the closure of $\mathcal{H}_8$ is also in P.

Corollary 6.27. $\mathsf{RSAT}(\widehat{\mathcal{H}}_8) \in \mathsf{P}$

Apart from the relations of $\mathcal{N}_1$ and $\mathcal{N}_2$, which cannot be included in any tractable subset of RCC-8 that contains all base relations (see Lemma 6.11), the only relations not contained in $\widehat{\mathcal{H}}_8$ are those that contain EQ and NTPP but not TPP, and the same for the converse relations.

Theorem 6.28. $\widehat{\mathcal{H}}_8$ *contains the following 148 RCC-8 relations:*

$$\widehat{\mathcal{H}}_8 = \text{RCC-8} \setminus (\mathcal{N}_1 \cup \mathcal{N}_2 \cup \mathcal{N}_3)$$

where

$$\mathcal{N}_3 = \{R \mid \{\mathsf{EQ}\} \subseteq R \text{ and } ((\{\mathsf{NTPP}\} \subseteq R, \{\mathsf{TPP}\} \not\subseteq R) \\ \text{ or } (\{\mathsf{NTPP}^{-1}\} \subseteq R, \{\mathsf{TPP}^{-1}\} \not\subseteq R))\}$$

and $\mathcal{N}_1$ and $\mathcal{N}_2$ were defined in Lemma 6.11.

6.3.2 Maximality of $\widehat{\mathcal{H}}_8$ with Respect to Tractability

Our goal is to find maximal tractable subsets of RCC-8 since these subsets mark the boundary between tractability and NP-hardness, i.e., any subset of one of these sets is tractable, any superset is NP-hard. For proving that $\widehat{\mathcal{H}}_8$ is a maximal tractable subset of RCC-8 we have to show that no relation of $\mathcal{N}_3$ can be added to $\widehat{\mathcal{H}}_8$ without making RSAT intractable. By computing the closure of $\widehat{\mathcal{H}}_8$ with each relation of $\mathcal{N}_3$ we get the following lemma.

Lemma 6.29. *The closure of every subset of* RCC-8 *containing $\widehat{\mathcal{H}}_8$ and one relation of $\mathcal{N}_3$ contains the relation $\{\mathsf{EQ}, \mathsf{NTPP}\}$.*

Therefore it is sufficient to prove NP-hardness of $\mathsf{RSAT}(\widehat{\mathcal{H}}_8 \cup \{\{\mathsf{EQ}, \mathsf{NTPP}\}\})$ for showing that $\widehat{\mathcal{H}}_8$ is a maximal tractable subset of RCC-8.

Theorem 6.30. $\mathsf{RSAT}(\widehat{\mathcal{H}}_8 \cup \{\{\mathsf{EQ}, \mathsf{NTPP}\}\})$ *is NP-complete.*

Proof. Transformation of 3SAT to $\mathsf{RSAT}(\widehat{\mathcal{H}}_8 \cup \{\{\mathsf{EQ}, \mathsf{NTPP}\}\})$.[7] Let $\mathcal{V} = \{v_1, v_2, \ldots, v_n\}$ be a set of variables and $\mathcal{C} = \{c_1, c_2, \ldots, c_m\}$ be a set of clauses of an arbitrary instance of 3SAT with $c_i = \{l_{i,1}, l_{i,2}, l_{i,3}\}$, where $l_{i,j}$ are literals over variables of $\mathcal{V}$. We will construct a set of spatial constraints Θ using only relations of $\widehat{\mathcal{H}}_8 \cup \{\{\mathsf{EQ}, \mathsf{NTPP}\}\}$, such that Θ is satisfiable if and only if $\mathcal{C}$ is a positive instance of 3SAT, using the following three transformation steps:

---

[7] The structure of this proof parallels the proof of Theorem 6.3 in Section 6.1. Here, however, $R_t = \mathsf{NTPP}$ and $R_f = \mathsf{EQ}$. The comments on "polarity constraints" and "clause constraints" given in Section 6.1 hold accordingly.

(1) For each variable $v_L \in \mathcal{V}$ the spatial variables x_L, y_L, $x_{\neg L}$, and $y_{\neg L}$ are introduced by adding the spatial constraints $x_L\{EQ, NTPP\}y_L$ and $x_{\neg L}\{EQ, NTPP\}y_{\neg L}$ to Θ. Additionally, the following polarity constraints are added to Θ (see Figure 6.10):

$$x_L\{EC, NTPP\}x_{\neg L}, y_L\{TPP\}y_{\neg L},$$
$$x_L\{TPP, NTPP\}y_{\neg L}, y_L\{EC, TPP\}x_{\neg L}.$$

(2) For each literal occurrence $l_{i,j}$ the spatial variables $x_{i,j}$ and $y_{i,j}$ are introduced by adding the spatial constraint $x_{i,j}\{EQ, NTPP\}y_{i,j}$ to Θ. Depending on whether the literal occurrence is positive or negative different polarity constraints have to be added to Θ.

a) $l_{i,j} \equiv v_L$:

$$x_{i,j}\{EC, NTPP\}x_{\neg L}, y_{i,j}\{TPP\}y_{\neg L},$$
$$x_{i,j}\{TPP, NTPP\}y_{\neg L}, y_{i,j}\{EC, TPP\}x_{\neg L}.$$

b) $l_{i,j} \equiv \neg v_L$:

$$x_{i,j}\{EC, NTPP\}x_L, y_{i,j}\{TPP\}y_L,$$
$$x_{i,j}\{TPP, NTPP\}y_L, y_{i,j}\{EC, TPP\}x_L.$$

(3) For each clause $c_i = \{l_{i,1}, l_{i,2}, l_{i,3}\}$ the following clause constraints are added to Θ:

$$y_{i,1}\{NTPP^{-1}\}x_{i,2}, y_{i,2}\{NTPP^{-1}\}x_{i,3}, y_{i,3}\{NTPP^{-1}\}x_{i,1}.$$

With this transformation for every variable as well as for every literal occurrence two spatial variables x and y (with the appropriate indices) are introduced. When a literal occurrence or a variable is assigned *true*, the corresponding spatial variables hold the relation $x\{NTPP\}y$. When a literal occurrence or a variable is assigned *false*, the corresponding spatial variables hold the relation $x\{EQ\}y$.

Transformation step (1) introduces the spatial variables corresponding to the positive and the negative literal of each variable. The polarity constraints assure that both literals have complementary assignments (see Figure 6.10). Transformation step (2) introduces spatial variables for every literal occurrence. Again, the polarity constraints assure correct assignments. Finally, transformation step (3) makes sure that at least one literal occurrence of every clause is *true*. If all literal occurrences of a clause are *false*, the corresponding spatial variables hold the relation $\{EQ\}$. Then there is a path starting at $x_{i,1}$, passing $x_{i,2}$ and $x_{i,3}$, and ending at $x_{i,1}$ where $\{NTPP^{-1}\}$ and $\{EQ\}$ are the only occurring relations, which is inconsistent. All other combinations are possible. We now have to show that an instance of 3SAT

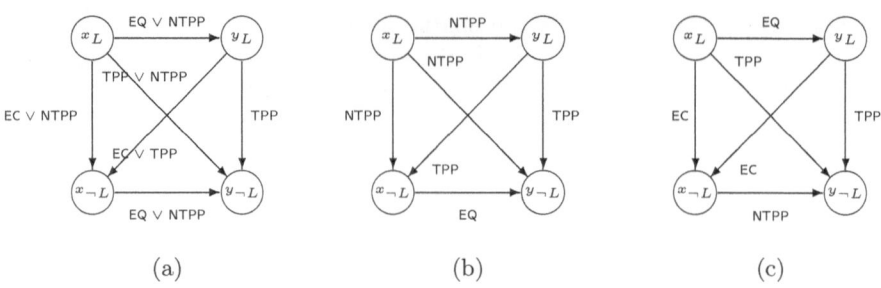

Fig. 6.10. The polarity constraints (a) for the transformation of 3SAT assure that positive and negative literals of the same variable have opposite assignments: (b) and (c) are the only consistent refinements of the relations to base relations

has a solution if and only if the set of spatial constraints Θ obtained by the given transformation is consistent.

RSAT $\Rightarrow$ 3SAT: Suppose that the set of spatial constraints Θ obtained by transformation from a given instance Σ of 3SAT is consistent, and suppose that θ is a consistent instantiation of Θ. Then an assignment σ that satisfies Σ can be obtained in the following way: For every variable $v_L \in \mathcal{V}$, if $\theta(x_L)\{\text{NTPP}\}\theta(y_L)$ holds, then $\sigma(v_L)$ is *true*, otherwise $\sigma(v_L)$ is *false*.

3SAT $\Rightarrow$ RSAT: Suppose that Σ is a positive instance of 3SAT and suppose that σ is an assignment that satisfies Σ. Then the set of spatial constraints Θ obtained by transformation from Σ with respect to σ is consistent. We will show this by constructing a spatial configuration that holds all relations of Θ.

First we will point out some properties of Θ. For every literal and every literal occurrence that is assigned *false*, the corresponding spatial variables hold the relation **EQ** and can therefore be treated as a single spatial variable. Figure 6.11(a) shows the three spatial variables corresponding to a variable P with $\sigma(P) = true$ (placed inside the dashed box) and spatial variables corresponding to a positive and a negative literal occurrence of P. Figure 6.11(b) shows the same for a variable Q where $\sigma(Q) = false$. For each variable there might be different corresponding literal occurrences, but all hold the same constraints. The **NTPP** relations in Figure 6.11 pointing to or from nowhere indicate the clause constraints required by transformation step (3). Note that for each variable $V \in \mathcal{V}$ the spatial variable $y_{\neg V}$ contains all other spatial variables corresponding to V and all spatial variables corresponding to the literal occurrences of V.

For constructing a spatial configuration that holds all constraints of Θ we start with m different regions C_i, one for each clause $c_i \in \mathcal{C}$, such that $C_i\{\text{DC}\}C_j$ holds for every $i \neq j$. All regions corresponding to the literal occurrences of a clause c_i are placed within the region C_i. The three regions corresponding to each variable consist of different pieces where a piece

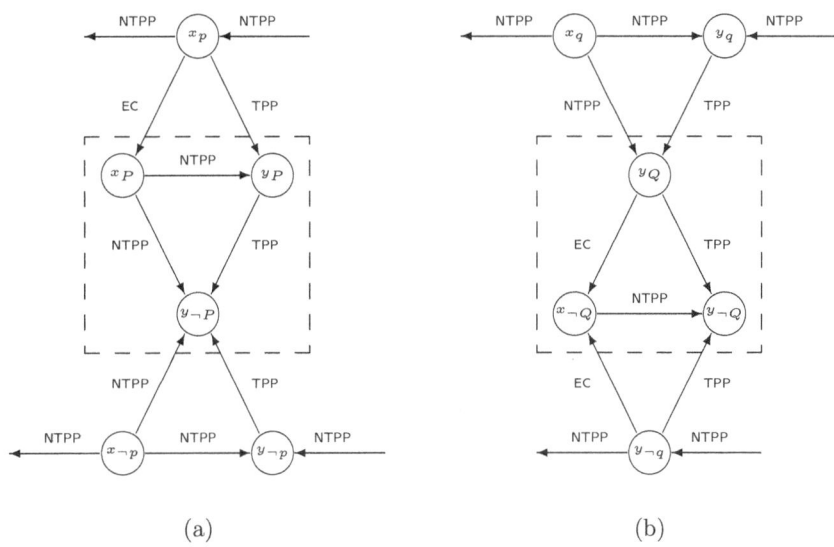

Fig. 6.11. Spatial variables corresponding to the variables P and Q and to their literal occurrences

is contained in C_i if a literal occurrence of the variable is contained in c_i. All regions within region C_i can be arranged consistently in a way that all required relations hold. Since the regions corresponding to the literal occurrences hold only the relations $\{TPP\}$, $\{NTPP\}$ or $\{EC\}$ with the pieces of the regions corresponding to the variables, they hold the same relations with the compound regions. All required relations hold for these regions, so we have a consistent model for Θ.

The specified transformation takes linear time in the number of clauses, so it has been proven that $\mathsf{RSAT}(\widehat{\mathcal{H}}_8 \cup \{\{EQ, NTPP\}\})$ is NP-hard. Because of Corollary 6.22 it is also NP-complete. ∎

Lemma 6.29 together with Theorem 6.30 results in the following theorem.

Theorem 6.31. $\widehat{\mathcal{H}}_8$ *is a maximal tractable subset of* RCC-8.

As it has not been proven that adding relations of $\mathcal{N}_3$ to the base relations results in intractability of RSAT, there might be other maximal tractable subsets of RCC-8 containing all base relations.

As $\widehat{\mathcal{H}}_8$ is a tractable subset of RCC-8, the intersection of RCC-5 and $\widehat{\mathcal{H}}_8$ is also tractable. We will call this subset $\widehat{\mathcal{H}}_5$.

Proposition 6.32. $\widehat{\mathcal{H}}_5$ *contains all relations of* RCC-5 *except for the relations* $\{PP, PP^{-1}\}$, $\{DR, PP, PP^{-1}\}$, $\{PP, PP^{-1}, EQ\}$ *and* $\{DR, PP, PP^{-1}, EQ\}$.

The following lemma can easily be obtained by computing the closure of the employed sets.

Lemma 6.33. *The closure of the set of all* RCC-5 *base relations together with one of the relations* $\{\mathsf{DR}, \mathsf{PP}, \mathsf{PP}^{-1}\}$, $\{\mathsf{PP}, \mathsf{PP}^{-1}, \mathsf{EQ}\}$ *or* $\{\mathsf{DR}, \mathsf{PP}, \mathsf{PP}^{-1}, \mathsf{EQ}\}$ *contains* $\{\mathsf{PP}, \mathsf{PP}^{-1}\}$.

Theorem 6.34. $\widehat{\mathcal{H}}_5$ *is the only maximal tractable subset of* RCC-5 *containing all base relations.*

Proof. $\widehat{\mathcal{H}}_5$ is by definition a tractable subset of RCC-5. To prove NP-hardness of RCC-5 the relations $\{\mathsf{PO}\}$, $\{\mathsf{PP}, \mathsf{PP}^{-1}\}$ and $\{*\}$ were used in Lemma 6.3. With Lemma 6.33 it follows that $\widehat{\mathcal{H}}_5$ contains all relations except for those making a set containing all base relations NP-complete.   ∎

6.4 Applicability of Path-Consistency

In the previous section we proved that $\widehat{\mathcal{H}}_8$ is a tractable subset of RCC-8. Thus, RSAT($\widehat{\mathcal{H}}_8$) can be decided in polynomial time. So far this can be done by first transforming a set of $\widehat{\mathcal{H}}_8$-constraints to a propositional Horn formula and then solving the resulting Horn formula in time linear in the number of literals [40]. Because the number of literals is of the order n^4, this way of solving RSAT does not appear to be very efficient.

The previously mentioned path-consistency method (see Section 2.4) with a running time of $O(n^3)$ is much easier to apply than the above described method, but it is not complete in general. In order to apply this simple and popular method also for deciding consistency of RSAT($\widehat{\mathcal{H}}_8$), we have to prove completeness of the path-consistency method for this task.

In this section we first prove that the path-consistency method is sufficient for deciding consistency of RSAT($\mathcal{H}_8$) and based on this that it is also sufficient for deciding consistency of RSAT($\widehat{\mathcal{H}}_8$). This is done by showing that the path-consistency method finds an inconsistency whenever positive unit resolution resolves the empty clause from the corresponding propositional formula. Positive unit resolution (PUR) is a resolution strategy in which in every resolution step at least one of the two resolved clauses is a positive unit clause, i.e., a clause containing a single positive literal. As PUR is refutation-complete for Horn formulas [86], it follows that the path-consistency method decides consistency of RSAT($\mathcal{H}_8$).

6.4.1 Applying Positive Unit Resolution to the Horn Clauses of RCC-8

The only way to derive the empty clause using PUR is resolving a positive and a negative unit clause of the same variable. Since the Horn formulas that are used contain only a few different types of clauses, there are only a few ways of deriving unit clauses using PUR. In this subsection we will show how unit clauses can be derived, and how this relates to the structure of the initial set of constraints. In the following we will first point out some important observations made from the transformation of a set of RCC-8 constraints to propositional logic as specified in Proposition 6.19 in Section 6.2. For this we need some definitions.

Definition 6.35.

- R_K denotes the set of all relations of $\mathcal{H}_8$ whose abbreviated form contains the conjunct K (see Appendix A.1). $R_K(x, y)$ means that the relation between x and y is one of R_K.

- $R_{K_1, K_2, \ldots, K_n}$ denotes $R_{K_1} \cup R_{K_2} \cup \cdots \cup R_{K_n}$

- $R_{\sigma*}$ is written instead of $R_{\sigma, \delta \vee \sigma, \delta i \vee \sigma}$ for any abbreviation σ.

- The clause $\{X_w^*\}$ denotes either one of the clauses $\{X_w\}$ or $\{X_w^i\}$.

Proposition 6.36. Let Θ be a set of $\mathcal{H}_8$-constraints and $c(\Theta)$ be the corresponding set of Horn clauses obtained by the transformation specified in Proposition 6.19. The following observations result from the transformation of the abbreviated form of Θ to $c(\Theta)$:

1. For every world w either one or two unit input clauses are contained in $c(\Theta)$ and at least one of these is positive.

2. If the unit clauses $\{X_w\}$ and $\{Y_w\}$ or $\{X_w^i\}$ and $\{Y_w^i\}$ are input clauses of $c(\Theta)$, then $R_{\neg\delta}(x, y) \in \Theta$ or $R_{\neg\delta i}(x, y) \in \Theta$, respectively.

3. If the unit clause $\{\neg X_w\}$ or $\{\neg X_w^i\}$ is an input clause of $c(\Theta)$, there must be a spatial variable $Y \in Var(\Theta)$ with $R_{\neg\gamma}(x, y) \in \Theta$ or $R_{\neg\gamma i}(x, y) \in \Theta$, respectively. In this case $\{Y_w\}$ is also an input clause of $c(\Theta)$.

4. If none of the unit clauses $\{X_w^*\}$ is input clause of $c(\Theta)$ and if one of them is derivable from $c(\Theta)$, then there must be a spatial variable $y \in Var(\Theta)$ such that $R_{\gamma*, \gamma i*}(x, y)$ holds and the corresponding unit clause $\{Y_w^*\}$ is present. The clauses corresponding to $R_{\gamma i*}$ can only be used to derive the unit clause $\{X_w^i\}$ but not the unit clause $\{X_w\}$.

5. If one of the unit clauses $\{\neg X_w^*\}$ is derivable from $c(\Theta)$, then there is a spatial variable $y \in Var(\Theta)$ such that $R_{\delta, \delta i}(x, y)$ holds and the corresponding unit clause $\{Y_w^*\}$ is present. The clauses corresponding to $R_{\delta i}$ can only be used to derive the unit clause $\{\neg X_w^i\}$ but not the unit clause $\{X_w\}$.

6. If $\{X_w^i\}$ is present, then $\{X_w\}$ can be derived from $c(CP)$.

7. If no positive unit clause $\{X_w^i\}$ is present for a world w, it can only be derived using a clause introduced by $R_{\gamma i}(x, y)$ for some spatial variable y or using $c(RP)$. In the latter case only the clause $\{X_w^{h(x)}\}$ can be derived.

The sets of relations used in Proposition 6.36 contain the following relations (see Appendix A.1):

$$
\begin{aligned}
R_\gamma &= \{\{\mathsf{TPP}^{-1}\}, \{\mathsf{NTPP}^{-1}\}, \{\mathsf{EQ}\}, \{\mathsf{TPP}^{-1}, \mathsf{NTPP}^{-1}\}, \\
&\quad \{\mathsf{TPP}^{-1}, \mathsf{NTPP}^{-1}, \mathsf{EQ}\}\}, \\
R_{\delta\vee\gamma} &= \{R \cup \{\mathsf{DC}\} | R \in R_\gamma\}, \\
R_{\delta\mathrm{i}\vee\gamma} &= \{R \cup \{\mathsf{EC}\} | R \in (R_\gamma \cup R_{\delta\vee\gamma})\}, \\
R_{\gamma\mathrm{i}} &= \{\{\mathsf{NTPP}^{-1}\}\}, \\
R_{\delta\vee\gamma\mathrm{i}} &= \{\{\mathsf{DC}, \mathsf{NTPP}^{-1}\}\}, \\
R_{\delta\mathrm{i}\vee\gamma\mathrm{i}} &= \{\{\mathsf{EC}, \mathsf{NTPP}^{-1}\}, \{\mathsf{DC}, \mathsf{EC}, \mathsf{NTPP}^{-1}\}\}, \\
R_\delta &= \{\{\mathsf{DC}\}\}, \\
R_{\delta\mathrm{i}} &= \{\{\mathsf{DC}\}, \{\mathsf{EC}\}, \{\mathsf{DC}, \mathsf{EC}\}\}.
\end{aligned}
$$

It can be seen, that $R_{\gamma\mathrm{i}} \subseteq R_\gamma$, $R_{\delta\vee\gamma\mathrm{i}} \subseteq R_{\delta\vee\gamma}$ and $R_{\delta\mathrm{i}\vee\gamma\mathrm{i}} \subseteq R_{\delta\mathrm{i}\vee\gamma}$, so, for example, $R_{\gamma*}$ can be written instead of $R_{\gamma*,\gamma\mathrm{i}*}$.

With PUR, positive unit clauses can only be derived in a very specific way which is based on the above observations. For this and for the rest of this section, the notion of "chains" will be central:

Definition 6.37 (R_K-chain).
An R_K-chain from X to Y, written as $R_K^*(x, y)$, is a sequence of constraints $R_K(x, z)$, $R_K(z, z')$, ... $R_K(z'', y)$.

Lemma 6.38. *Let the clauses $\{Y_w^*\}$ and $\{Z_w^*\}$ be input clauses of $c(\Theta)$ (if there is only one positive input clause for w then $y \equiv z$). A new positive unit clause $\{X_w^*\}$ can be derived from $c(\Theta)$ only if Θ contains an $R_{\gamma*}$-chain from x to y or an $R_{\gamma*}$-chain from x to z.*

Proof. According to Proposition 6.36, item (4), a positive unit clause can only be resolved if there is a spatial variable $z^1 \in Var(\Theta)$ such that $R_{\gamma*}(x, z^1)$ holds and the corresponding unit clause $\{Z_w^1\}$ is present. If $z^1 \not\equiv y$ or $z^1 \not\equiv z$ this clause cannot be an input clause so there must be another spatial variable $z^2 \in Var(\Theta)$ such that $R_{\gamma*}(z^1, z^2)$ holds and the corresponding unit clause $\{Z_w^2\}$ is present. This goes on until there is a spatial variable z^n with $z^n \equiv y$ or $z^n \equiv z$, i.e., there is an $R_{\gamma*}$-chain from x to y or an $R_{\gamma*}$-chain from x to z. ∎

Applying PUR has some side-effects on the possible relations of Θ. In order to demonstrate this side-effect, consider the constraint $x\{\mathsf{DC}, \mathsf{TPP}^{-1}\}y$. The abbreviated form of the relation $R = \{\mathsf{DC}, \mathsf{TPP}^{-1}\}$ contains the conjunct $\delta\vee\gamma$ (see Appendix A.1), i.e., $R \in R_{\delta\vee\gamma} \subseteq R_{\gamma*}$. The clauses corresponding to this conjunct are $\{\neg X_u^*, \neg Y_u^*, X_v^*, \neg Y_v^*\}$ for all $u, v \in W_0$ and are compounded of the clauses corresponding to the abbreviations δ and γ (see Proposition 6.19). These clauses can be used to derive the positive unit clause $\{X_w^*\}$ for some w only if the positive unit clauses $\{Y_w^*\}$, $\{X_u^*\}$, and $\{Y_u^*\}$ are present for some u.

In this case, i.e., if it is possible to derive $\{X_w^*\}$ as described above, the clause $\{\neg X_u^*, \neg Y_u^*\}$ which corresponds to the abbreviation δ produces the empty clause. Thus, from the initial two possibilities δ or γ the first one becomes inconsistent and, since the relation $\{DC\} \in R_\delta$, the constraint $x\{TPP^{-1}\}y$ must hold. We describe this side-effect by the notion of "refinement by PUR":

Definition 6.39 (refinement by PUR).
A constraint $xRy \in \Theta$ is refined by PUR to a constraint $xR'y$, such that $R' \subset R$, if a clause corresponding to the constraint $xR''y$, such that $R'' = R \setminus R'$, can be used to produce the empty clause.

In the above example, $x\{DC, TPP^{-1}\}y$ is refined by PUR to $x\{TPP^{-1}\}y$.

Lemma 6.40. *If the positive unit clause $\{X_w^*\}$ is derived using PUR, then every constraint of the required R_{γ^*}–chain will be refined by PUR to a constraint in R_γ.*

Proof. Let y and z be two successive regions of the R_{γ^*}–chain, holding either $R_{\delta \vee \gamma}(y, z)$ or $R_{\delta i \vee \gamma}(y, z)$. Then the required unit clause $\{Y_w^*\}$ is derived from a clause of the type $\{Y_w^*, \neg Z_w^*, \neg Y_u^*, \neg Z_u^*\}$, which consists of two disjunctively connected parts, the δ or δi part $\{\neg Y_u^*, \neg Z_u^*\}$ and the γ part $\{Y_w^*, \neg Z_w^*\}$. For this resolution the unit clauses $\{Y_u^*\}$ and $\{Z_u^*\}$ are necessary which are inconsistent with the clauses corresponding to $R_\delta(y, z)$ and $R_{\delta i}(y, z)$. ∎

In order to derive a positive unit clause from a particular R_{γ^*}-chain, other positive unit clauses are necessary for which other R_{γ^*}-chains might be required. In order to refer to all R_{γ^*}-chains that are used to derive a particular positive unit clause we introduce the notion of chain structure.

Definition 6.41 (chain structure).
Let xRy with $R \in R_{\delta \vee \gamma, \delta i \vee \gamma}$ be a constraint of an R_{γ^*}-chain. The chain structure of xRy contains all R_{γ^*}-chains used to refine xRy by PUR to a constraint in R_γ.

6.4.2 Relating Positive Unit Resolution to Path-Consistency

In this subsection we prove that the path-consistency method is sufficient for deciding consistency of $RSAT(\mathcal{H}_8)$, by showing that for every set Θ of constraints over $\mathcal{H}_8$ whenever PUR produces the empty clause, the path-consistency method finds an inconsistency in Θ. In order to relate PUR to the path-consistency method, we first show that if a constraint $xRy \in \Theta$ with $R \in R_{\gamma^*}$ is refined by PUR to a constraint in R_γ, then the path-consistency method applied to Θ also refines the constraint to a constraint in R_γ. This is proven by Noetherian induction on chain structures, which is defined on well-founded relations (see e.g. [164]).

Definition 6.42 (well-founded relation).
A relation $\prec$ on a set M is well-founded, if and only if every non empty subset of M has a minimal element, i.e. $\forall S \subseteq M.(S \neq \emptyset \rightarrow \exists x \in S.(\neg \exists y \in S.y \prec x))$.

Theorem 6.43 (Noetherian Induction). *Let $\prec$ be a well-founded relation on a set M. To prove a property P for all $x \in M$, it suffices to prove:*

$$\forall x \in M.\,(\forall y \in M.\,(y \prec x) \rightarrow P(y)) \rightarrow P(x)$$

Proof. Suppose that the set $A \subseteq M$ of elements not satisfying P is not empty. Then A has a minimal element m, i.e. $P(y)$ holds for all $y \prec m$. Then $P(m)$ also holds, which contradicts the assumption. So A must be empty, i.e. P holds for all $x \in M$. ∎

Before applying Noetherian induction to chain structures, we have to define a relation on chain structures and show that this relation is well-founded.

Definition 6.44 (relation on chain structures).
Let $\prec$ be a relation on chain structures, let S_1 be the chain structure of the constraint $x^1 R_1 y^1$ and let S_2 be the chain structure of the constraint $x^2 R_2 y^2$. $S_1 \prec S_2$ holds if and only if the constraint $x^1 R_1 y^1$ occurs in S_2.

Lemma 6.45. $\prec$ *is a well-founded relation.*

Proof. By definition, if S is a chain structure where only constraints in R_γ occur, then there is no chain structure S' with $S' \prec S$. Suppose that S is the chain structure of a constraint xRy and the same constraint is contained in S. If the occurrence of xRy is recursive in S, then S cannot be used to refine xRy to a constraint in R_γ, so S is no chain structure in our sense. If the occurrence of xRy is not recursive in S, then S can be replaced by S', the chain structure belonging to the last occurrence of xRy in S. There is only a finite number of regions, so every non empty set of chain structures has at least one minimal element. ∎

We have to prove that whenever a constraint xRy of an R_{γ^*}-chain is refined by PUR to a constraint in R_γ in the sense of Lemma 6.40, then the path consistency method also results in R_γ, i.e., the base relations $\{DC\}$ and $\{EC\}$ will be excluded from R. For this proof we need the following operations which can be verified using Table 4.2.

Proposition 6.46. *Let R be a relation of* RCC-8.

1. *R_γ is closed under composition.*

2. *$R_{\gamma i} \circ R_\gamma = R_{\gamma i}$ and $R_\gamma \circ R_{\gamma i} = R_{\gamma i}$*

3. *If $\{DC, EC\} \cap R = \emptyset$, then $\{DC, EC\} \cap (R_\gamma \circ R) = \emptyset$ and $\{DC, EC\} \cap (R \circ R_\gamma^{\smile}) = \emptyset$.*

4. $\{\mathsf{DC}, \mathsf{EC}\} \cap (R_\gamma \circ R_\gamma^\smile) = \emptyset$.

5. $\{\mathsf{DC}\} \cap (R_\gamma \circ \{\mathsf{EC}\} \circ R_\gamma^\smile) = \emptyset$.

6. $\{\mathsf{DC}, \mathsf{EC}\} \cap (R_{\gamma i} \circ \{\mathsf{EC}\} \circ R_\gamma^\smile) = \{\mathsf{DC}, \mathsf{EC}\} \cap (R_\gamma \circ \{\mathsf{EC}\} \circ R_{\gamma i}^\smile) = \emptyset$

Lemma 6.47. *If every constraint of an R_{γ^*}-chain K from x^1 to y^1 in Θ is refined by PUR to a constraint in R_γ, the path-consistency method applied to Θ results in $R_\gamma(x^1, y^1)$.*

Proof. Let $P(S) \equiv$ "If a constraint $xRy \in \Theta$, such that $R \in R_{\gamma^*}$, is refined by PUR to a constraint in R_γ using the chain structure S, then the path-consistency method applied to Θ also refines xRy to a constraint in R_γ". We will prove $P(S)$ with Noetherian induction.

Induction hypothesis: $\forall S' \prec S.\ \ P(S')$.

Suppose that the constraint xRy is refined by PUR to one of R_γ in order to obtain the clause $\{X_w^*\}$. If R is already in R_γ nothing has to be proven. If R is one of $R_{\delta \vee \gamma, \delta i \vee \gamma}$, the clause $\{X_w^*, \neg Y_w^*, \neg X_u^*, \neg Y_u^*\}$ is input clause of $c(\Theta)$. Since $R_{\delta, \delta i}(x, y)$ can be excluded with PUR, the unit clauses $\{X_u^*\}$ and $\{Y_u^*\}$ for some u must be present. Let the clauses $\{\tilde{x}_u^*\}$ and $\{\tilde{y}_u^*\}$ be the only unit input clauses of u, which can only be introduced by $R_{\neg \delta}(\tilde{x}, \tilde{y})$ or by $R_{\neg \delta i}(\tilde{x}, \tilde{y})$. (If $\tilde{x} \equiv \tilde{y}$ then the unit clause can be introduced by some other constraint which is not important in this analysis.) Then there are R_{γ^*}-chains from x to $\tilde{x}$ and from y to $\tilde{y}$ that are part of S. Six different cases, shown in Figure 6.12, must be distinguished:

Case 1: A chain structure $S' \prec S$ belongs to every constraint in $R_{\delta \vee \gamma, \delta i \vee \gamma}$ of the R_{γ^*}-chains from x to $\tilde{x}$ and from y to $\tilde{y}$. Since $P(S')$ holds by induction hypothesis, and R_γ is closed under composition, $R_\gamma(x, \tilde{x})$ and $R_\gamma(y, \tilde{y})$ are obtained by the path-consistency method.
  a. If $R \in R_{\delta \vee \gamma}$, then R is refined to one of R_γ by the path-consistency method because of items (3) and (5) of Proposition 6.46.
  b. If $R \in R_{\delta i \vee \gamma}$ then $\{X_w^i\}$ and $\{Y_w^i\}$ must be obtained, so the R_{γ^*}-chains from x to $\tilde{x}$ and from y to $\tilde{y}$ must each contain at least one relation of $R_{\gamma i}$ (see Proposition 6.36 item (7)). The two clauses cannot be obtained by $c(\mathsf{CP})$ because in this case they cannot have the same index i. Because of item (2) of Proposition 6.46 the path-consistency method results in $R_{\gamma i}(x, \tilde{x})$ and $R_{\gamma i}(y, \tilde{y})$. Because of items (3) and (6) of Proposition 6.46 it results in $R_\gamma(x, y)$.

The remaining five cases can be handled in the same way as case 1. All the R_{γ^*}-chains can be reduced to R_γ by applying the induction hypothesis. With the operations of Proposition 6.46 the path-consistency method reduces R to one of R_γ.

By Noetherian induction we proved that $P(S)$ holds for all chain structures S. Every relation of the R_{γ^*}-chain K can now be reduced to one of R_γ

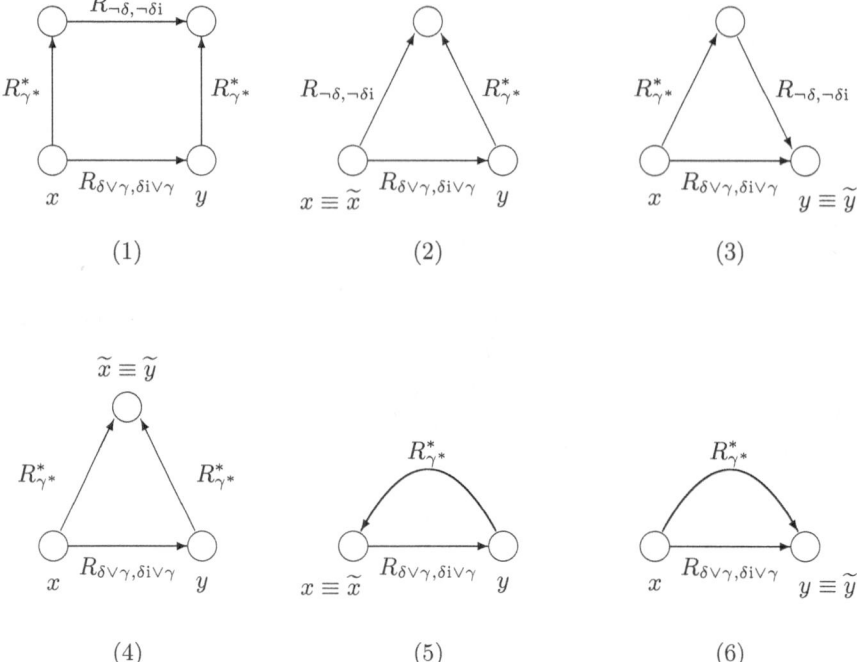

Fig. 6.12. The six possible cases of the proof to Lemma 6.47. In the lower row $\tilde{x}$ is equal to $\tilde{y}$

with the path-consistency method. Because R_γ is closed under composition, the path-consistency method results in $R_\gamma(x^1, y^1)$.   ∎

Using this lemma, we can now prove that the path-consistency method decides RSAT($\mathcal{H}_8$) by showing that whenever the empty clause can be derived, the path-consistency method results in an inconsistency.

Theorem 6.48. *The path-consistency method decides* RSAT($\mathcal{H}_8$).

Proof. Let Θ be an inconsistent set of $\mathcal{H}_8$-constraints and $c(\Theta)$ the corresponding set of Horn clauses obtained by the transformation specified in Proposition 6.19. Since Θ is inconsistent, the empty clause can be derived from $c(\Theta)$ using PUR. There are different possibilities of deriving the empty clause.

1. The empty clause is derived from $\{X_w\}$ and $\{\neg X_w\}$.
   a) $\{\neg X_w\}$ is input clause of $c(\Theta)$. Then, with Proposition 6.36 item (3), $\{Y_w\}$ is also an input clause for some spatial variable y with $R_{\neg\gamma}(x, y)$. $\{X_w\}$ is derived from $c(\Theta)$, so there must be an R_{γ^*}-chain

from x to y. Because of Lemma 6.47, the path-consistency method results in $R_\gamma(x, y)$, which is a contradiction to $R_{\neg\gamma}(x, y)$.

b) $\{X_w\}$ is the only positive input clause of w in $c(\Theta)$. Then $\{\neg X_w\}$ is derived from $c(\Theta)$. According to Proposition 6.36 item (5), there must be a spatial variable y with $R_{\delta,\delta i}(x, y)$ and $\{Y_w\}$ must be present. $\{Y_w\}$ is derived using an R_{γ^*}-chain from y to x, which is reduced to $R_\gamma(y, x)$ by the path-consistency method. This is a contradiction to $R_{\delta,\delta i}(x, y)$.

c) $\{X_w\}$ is an input clause of $c(\Theta)$ and $\{Z_w\}$ is also an input clause for some spatial variable z. According to Proposition 6.36 item (2), $R_{\neg\delta}(x, z)$ must hold. $\{\neg X_w\}$ is derived from $c(\Theta)$, so there must be a spatial variable y with $R_{\delta,\delta i}(x, y)$ (see Proposition 6.36 item (5)) and $\{Y_w\}$ must also be derived from $c(\Theta)$. So there is either an R_{γ^*}-chain from y to x, which is equal to (b), or an R_{γ^*}-chain from y to z which is refined to $R_\gamma(y, z)$ by the path-consistency method. If $R_\delta(x, y)$ holds, the path-consistency method results in an inconsistency (see Table 4.2). If $R_{\delta i}(x, y)$ holds, then one relation of the R_{γ^*}-chain from y to z must be from $R_{\gamma i}$, so the path consistency method results in $R_{\gamma i}(y, z)$, which is also inconsistent (see Table 4.2).

d) Neither $\{X_w\}$ nor $\{\neg X_w\}$ are input clauses of $c(\Theta)$, so both are derived using PUR. The negative clause can only be derived if there is a spatial variable y with $R_{\delta,\delta i}(x, y)$ (see Proposition 6.36 item (5)) and $\{Y_w\}$ can also be derived. As it was shown in the six cases of Lemma 6.47, $\{X_w\}$ and $\{Y_w\}$ can only be derived from $c(\Theta)$ if the path-consistency method results in an inconsistency with $R_{\delta,\delta i}(x, y)$.

2. The empty clause is derived from $\{X_w^i\}$ and $\{\neg X_w^i\}$. Only $\{X_w^i\}$ or $\{\neg X_w^{h(\neg x)}\}$ can be input clauses of $c(\Theta)$.

a) The empty clause is derived from $\{X_w^{h(\neg x)}\}$ and the input clause $\{\neg X_w^{h(\neg x)}\}$. According to Proposition 6.36 item (3) there is a spatial variable y with $R_{\neg\gamma i}(x, y)$ and $\{Y_w\}$ is also input clause. The clause $\{X_w^{h(\neg x)}\}$ is derived using an R_{γ^*}-chain from x to y. Because of Proposition 6.36 item (7), one of the relations of the chain must be in $R_{\gamma i}$, so the path consistency method results in $R_{\gamma i}(x, y)$ which contradicts $R_{\neg\gamma i}(x, y)$.

b) $\{X_w^i\}$ is input clause, then there is a spatial variable y with $R_{\neg\delta i}(x, y)$ (Proposition 6.36 item (2)) and $\{Y_w^i\}$ is also an input clause of $c(\Theta)$. In order to derive $\{\neg X_w^i\}$, there must be a spatial variable z with $R_{\delta,\delta i}(x, z)$ (Proposition 6.36 item (5)) and $\{Z_w^i\}$ must be present. $\{Z_w^i\}$ can be derived using an R_{γ^*}-chain from y to z. The path-consistency method results in $R_\gamma(y, z)$ which is, composed with $R_{\neg\delta i}(x, y)$, inconsistent with $R_{\delta,\delta i}(x, z)$ (see Table 4.2).

c) Neither $\{X_w^i\}$ nor $\{\neg X_w^i\}$ are input clauses and must be derived from
$c(\Theta)$. Then there must be a spatial variable y with $R_{\delta,\delta i}(x, y)$ (Proposition 6.36 item (2)) and $\{Y_w^i\}$ must be present. In order to obtain $\{X_w^i\}$ and $\{Y_w^i\}$ the six cases of Lemma 6.47 must be considered, where $R_{\delta,\delta i}(x, y)$ holds instead of $R_{\gamma*}(x, y)$. Since $R_{\delta,\delta i}$ could be excluded in Lemma 6.47, it can also be excluded now, so the path-consistency method finds the inconsistency.

Thus, whenever PUR derives the empty clause, an inconsistency is found by the path-consistency method. Since positive unit resolution is refutation-complete for propositional Horn formulas, the path-consistency method is sufficient for deciding consistency of $\mathsf{RSAT}(\mathcal{H}_8)$. ∎

6.4.3 Path-Consistency for the Full Set of Tractable Relations

The path-consistency method can be used to decide $\mathsf{RSAT}(\widehat{\mathcal{H}}_8)$ when every constraint of $\widehat{\mathcal{H}}_8 \setminus \mathcal{H}_8$ is transformed to constraints in $\mathcal{H}_8$ according to Theorem 6.5. As this way of deciding $\mathsf{RSAT}(\widehat{\mathcal{H}}_8)$ is pretty awkward, we will prove that the path-consistency method can decide $\mathsf{RSAT}(\widehat{\mathcal{H}}_8)$ directly without any preprocessing. This will be shown in the same way as it was shown for $\mathcal{H}_8$, namely, that the path-consistency method finds an inconsistency whenever PUR derives the empty clause. For this we have to transform constraints in $\widehat{\mathcal{H}}_8 \setminus \mathcal{H}_8$ to propositional Horn formulas.

Proposition 6.49. *Let S be the set of* RCC-8 *relations transformable to a propositional Horn formula. Then every constraint xSy with $S \in \mathcal{S}$ can be transformed to the propositional Horn formula $p'(S(x, y))$. If $S \in \mathcal{H}_8$, then $p'(S(x, y)) = p(m(S(x, y)))$ where p is the transformation given in Proposition 6.19. Since any relation of $\widehat{\mathcal{H}}_8 \setminus \mathcal{H}_8$ can be obtained by composition, converse or intersection of relations of $\mathcal{H}_8$, the propositional Horn formula of xRy where $R \in \widehat{\mathcal{H}}_8 \setminus \mathcal{H}_8$ can be obtained using the following construction inductively:*

1. *If $R = S \circ T$, where $S, T \in \mathcal{S}$ and $R \notin \mathcal{S}$, then introduce a pseudo variable z which is only related to the spatial variables x and y, holding xSz and zTy. Therefore $p'(xRy) = p'(xSz) \wedge p'(zTy)$. Add R to $\mathcal{S}$.*

2. *If $R = S \cap T$, where $S, T \in \mathcal{S}$ and $R \notin \mathcal{S}$, then $p'(xRy) = p'(xSy) \wedge p'(xTy)$. Add R to $\mathcal{S}$.*

3. *If $R = S^{\smile}$, where $S \in \mathcal{S}$ and $R \notin \mathcal{S}$, then $p'(xRy) = p'(ySx)$. Add R to $\mathcal{S}$.*

For proving that the path-consistency method decides $\mathsf{RSAT}(\mathcal{H}_8)$, the $R_{\gamma*}$-chain was the central part as it is the only way to derive a positive unit clause. Some relations of $\widehat{\mathcal{H}}_8 \setminus \mathcal{H}_8$ can be constructed using relations of $R_{\gamma*}$, so these relations can also be used to derive positive unit clauses. As all of these relation must be analyzed separately, we try to keep their number as small as possible. Therefore, we consider only relations of $\widehat{\mathcal{H}}_8 \setminus \mathcal{H}_8$ that cannot be constructed without using relations of R_{γ^*}. In Appendix A.2 we give a list of all relations of $\widehat{\mathcal{H}}_8 \setminus \mathcal{H}_8$ and how they can be functionally constructed from $\mathcal{H}_8$-relations. We have chosen a construction of all relations such that the following analysis is as simple as possible. The set of relations that can be used to derive the empty clause will be denoted by R_e.

Lemma 6.50. *Additional to R_{γ^*}, R_e contains the following relations:*

$$\{DC, EC, PO, NTPP, TPP^{-1}, NTPP^{-1}\} = \{EC, NTPP\} \circ \{DC, EQ\},$$
$$\{DC, EC, PO, TPP, NTPP, NTPP^{-1}\} = \{DC, EQ\} \circ \{EC, NTPP^{-1}\},$$
$$\{EC, PO, EQ, TPP^{-1}, NTPP^{-1}\} = \{EQ, TPP^{-1}, NTPP^{-1}\} \circ \{EC, EQ\},$$
$$\{DC, EC, PO, EQ, TPP^{-1}, NTPP^{-1}\} = \{EC, EQ\} \circ \{DC, EQ\},$$

as well as the intersection of these relations with other relations of $\widehat{\mathcal{H}}_8$.

Proof. This can be proven by computing the closure of $\mathcal{H}_8 \setminus R_{\gamma^*}$ under composition and intersection. All relations not contained in this closure can be constructed by intersection of these relations with either the four specified relations or the relations of R_{γ^*}. ∎

Instead of R_{γ^*}-chains we now consider R_e-chains which contain constraints over R_e. Similar to Definition 6.41 and Definition 6.44, we define *extended chain structures* on constraints over R_e and a well-founded relation on extended chain structures.

Definition 6.51 (extended chain structure).

1. *The* extended chain structure *of xRy with $R \in R_e$ contains all R_e-chains used to refine R to a relation of R_γ.*

2. *Let $\prec_e$ be a relation on extended chain structures, let T_1 be the extended chain structure of the constraint $x^1 R_1 y^1$ and T_2 be the extended chain structure of the constraint $x^2 R_2 y^2$, where $R_1, R_2 \in R_e$. $T_1 \prec_e T_2$ holds if and only if $x^1 R_1 y^1$ occurs in T_2.*

Analogous to Lemma 6.45 it can be proven that $\prec_e$ is well-founded. Similar to Lemma 6.47 we can relate positive unit resolution and path-consistency also for relations of R_e.

Lemma 6.52. *If every constraint of an R_e-chain K from X^1 to Y^1 in Θ is refined by PUR to a constraint in R_γ, the path-consistency method applied to Θ results in $R_\gamma(X^1, Y^1)$.*

Proof. $P(T) \equiv$ "If a constraint $xRy \in \Theta$, such that $R \in R_e$, is refined by PUR to a constraint in R_γ using the chain structure T, then the path-consistency method applied to Θ also refines xRy to a constraint in R_γ". We will prove $P(T)$ by Noetherian induction.

Induction hypothesis: $\forall T' \prec T. \quad P(T')$.

Suppose that the constraint xRy, with $R \in R_e$, is refined by PUR to a constraint in R_γ in order to obtain the clause $\{X_w^*\}$. We have to distinguish different cases for R.

1. $R \in R_{\gamma^*}$: The six different cases of Lemma 6.47 can also be applied here. By applying the induction hypothesis, all R_e-chains of the extended chain structure of xRy will be refined to R_γ with the path-consistency method. Therefore R will also be refined to R_γ with the path-consistency method.

2. $R = \{DC, EC, PO, NTPP, TPP^{-1}, NTPP^{-1}\}$: Only the second relation of the construction of R is in R_{γ^*}, so $\{X_w^*\}$ cannot be obtained with this relation.

3. $R = \{DC, EC, PO, TPP, NTPP, NTPP^{-1}\}$: Suppose this relation holds between the spatial variables x and y using the pseudo variable z. Then clauses of the type $\{\neg X_a^*, \neg Z_a^*, \neg X_b^*, Z_b^*\}$, $\{\neg X_a^*, \neg Z_a^*, X_b^*, \neg Z_b^*\}$, and $\{\neg Z_a^*, \neg Y_a^*, Z_b^*, \neg Y_b^*\}$ for all $a, b \in W_0$ as well as $\{Z_c\}$, $\{Y_c\}$, $\{Z_d\}$ and $\{\neg Y_d\}$ for some $c, d \in W_0$ are input clauses. In order to derive $\{X_w^*\}$, the unit clauses $\{Z_w^*\}$, $\{X_u^*\}$, and $\{Z_u^*\}$ are necessary for some u. There are different possibilities of how $\{Z_u^*\}$ can be derived:

   a) If $\{Z_u\}$ is the only positive input clause of u, then $\{X_u\}$ is derived using $\{Z_v^*\}$ and $\{X_v^*\}$ for some $v \neq u$. Then $\{X_w^*\}$ can also be derived using $\{Z_v^*\}$ and $\{X_v^*\}$, which contradicts the assumption that it is derived using $\{Z_u\}$ and $\{X_u\}$.

   b) If $\{Z_u\}$ and $\{Y_u\}$ are input clauses, $\{X_u\}$ can either be derived as described in (a) which results in a contradiction, or there might be an additional R_e-chain from x to y, not passing z. By applying the induction hypothesis, $R_\gamma(x, y)$ is obtained with the path-consistency method, if it is refined by PUR to the same constraint.

   c) If $\{Z_u^*\}$ is not an input clause, there must be a spatial variable $\tilde{z}_u$ that introduces u. If the R_e-chain from z to $\tilde{z}_u$ passes x, then the clauses $\{Z_v^*\}$ and $\{X_v^*\}$ for some $v \neq u$ are necessary. This is again a contradiction to our assumption. If the chain passes y, then the clauses $\{Y_u^*\}$, $\{Z_v^*\}$ and $\{Y_v^*\}$ for some v are necessary. In order to derive $\{Z_v^*\}$, the clauses $\{Y_v^*\}$, $\{Z_s^*\}$, and $\{Y_s^*\}$ for some $s \neq v$ are necessary. Then $\{Z_u^*\}$ can also be derived using $\{Z_s^*\}$ and $\{Y_s^*\}$, which contradicts the assumption that it is derived using $\{Z_v^*\}$ and $\{Y_v^*\}$.

4. $R = \{EC, PO, EQ, TPP^{-1}, NTPP^{-1}\}$: Suppose this relation holds between the spatial variables x and y using the pseudo variable z. Then clauses of the type $\{X_a^*, \neg Z_a^*\}$, $\{\neg Z_a^*, \neg Y_a^*, \neg Z_b^*, Y_b^*\}$, $\{\neg Z_a^*, \neg Y_a^*, Z_b^*, \neg Y_b^*\}$ for all $a, b \in W_0$ as well as $\{X_c\}$, $\{Z_c\}$, $\{Z_d\}$, and $\{Y_d\}$ for some $c, d \in W_0$ are input clauses. In order to derive $\{X_w^*\}$, the unit clause $\{Z_w^*\}$ is necessary. If $\{Z_w\}$ is an input clause, the relation between x and z is not refined by PUR. If $\{Z_w^*\}$ is not an input clause, the clauses $\{Y_w^*\}$, $\{Z_u^*\}$, and $\{Y_u^*\}$

for some u are necessary. In order to derive $\{Z_u^*\}$ the clauses $\{Y_u^*\}$, $\{Z_v^*\}$ and $\{Y_v^*\}$ for some $v \neq u$ are necessary. Then $\{Z_w^*\}$ can also be derived using $\{Z_v^*\}$ and $\{Y_v^*\}$, which contradicts the assumption that it is derived using $\{Z_u^*\}$ and $\{Y_u^*\}$.

5. $R = \{DC, EC, PO, EQ, TPP^{-1}, NTPP^{-1}\}$: Suppose this relation holds between the spatial variables x and y using the pseudo variable z. Then clauses of the type $\{\neg X_a^*, \neg Z_a^*, \neg X_b^*, Z_b^*\}$, $\{\neg X_a^*, \neg Z_a^*, X_b^*, \neg Z_b^*\}$, $\{\neg Z_a^*, \neg Y_a^*, \neg Z_b^*, Y_b^*\}$, and $\{\neg Z_a^*, \neg Y_a^*, Z_b^*, \neg Y_b^*\}$ for all $a, b \in W_0$, as well as $\{X_c\}$ and $\{Z_c\}$ for some $c \in W_0$ are input clauses. In order to derive $\{X_w^*\}$, the unit clauses $\{Z_w^*\}$, $\{X_u^*\}$, and $\{Z_u^*\}$ are necessary for some u. $\{Z_u^*\}$ can be obtained with an R_e-chain from z to $\tilde{z}_u$ passing either x or y. If it passes x, the clauses $\{X_v^*\}$, and $\{Z_v^*\}$ for some $v \neq u$ are necessary, which is a contradiction. If it passes y, the clauses $\{Y_u^*\}$, $\{Y_v^*\}$ and $\{Z_v^*\}$ for some $v \neq u$ are necessary. $\{Z_v^*\}$ can be obtained in two ways:

   a) If $\{Z_v\}$ is an input clause, $\{Y_v\}$ can be derived using the clauses $\{Z_s^*\}$ and $\{Y_s^*\}$ for some s. Then $\{Z_u^*\}$ can also be derived using $\{Z_s^*\}$ and $\{Y_s^*\}$, which contradicts the assumption that it is derived using $\{Z_v^*\}$ and $\{Y_v^*\}$.

   b) If the chain for obtaining $\{Z_v^*\}$ passes x, the clauses $\{Z_s^*\}$ and $\{X_s^*\}$ for some s are necessary. If it passes y, the clauses $\{Z_s\}$ and $\{Y_s\}$ for some s are necessary. Both possibilities result in a contradiction to previous assumptions.

6. R is constructed by intersection: The intersection of two relations is a refinement of both relations, so if one of the relations is refined to one of R_γ, this also holds for the intersection of the two relations. Some relations are refinements of a relation of R_γ but are not part of R_γ. These relations can be treated as if they were part of R_γ, because the specified properties of Proposition 6.46 also hold for them.

As R_γ is closed under composition, the proof is completed.   ∎

We are now ready to prove the main theorem of this section, namely, that the path-consistency method is sufficient for deciding $RSAT(\widehat{\mathcal{H}}_8)$.

Theorem 6.53. *The path-consistency method decides* $RSAT(\widehat{\mathcal{H}}_8)$.

Proof. Let Θ be an inconsistent set of $\widehat{\mathcal{H}}_8$-constraints, $p'(\Theta)$ the equivalent propositional Horn formula as specified in Proposition 6.49 and $c'(\Theta)$ the corresponding set of Horn clauses. Since Θ is inconsistent, the empty clause can be derived from $c'(\Theta)$ using PUR. Suppose that the empty clause is derived by $\{X_w^*\}$ and $\{\neg X_w^*\}$. The same proof as in Theorem 6.48 can be applied here, when it is based on Lemma 6.52 instead of Lemma 6.47. The only difference is that x might be a pseudo variable. In this case $\{\neg X_w^*\}$ must be input clause, as it cannot be derived by PUR. For this there must either be a spatial variable y with $R_{\neg \pi}(y, x)$ or a spatial variable z with $R_{\neg \gamma, \neg \gamma i}(x, z)$.

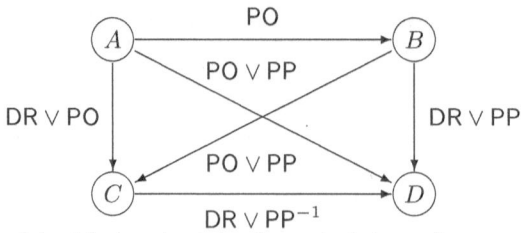

Fig. 6.13. Path consistent but not minimal constraint graph

As there is no such spatial variable, x cannot be a pseudo variable, and then the empty clause cannot be derived from $\{X_w^*\}$ and $\{\neg X_w^*\}$. ∎

This theorem can be easily transfered to RCC-5.

Corollary 6.54. *The path-consistency method decides* $\mathsf{RSAT}(\widehat{\mathcal{H}}_5)$.

Another interesting question is whether the path-consistency method also decides the minimal-label problem $\mathsf{RMIN}(\widehat{\mathcal{H}}_8)$. As the following proposition shows this is not the case even for the set $\widehat{\mathcal{H}}_5$.

Proposition 6.55. *The path-consistency method is not sufficient for solving the minimal-label problem* $\mathsf{RMIN}(\widehat{\mathcal{H}}_5)$.

Proof. Figure 6.13 shows a constraint graph that is path-consistent but not minimal. The relation between A and D can be refined to PO but not to PP. ∎

Since RMIN and RSAT are equivalent under polynomial Turing reductions, $\mathsf{RMIN}(\widehat{\mathcal{H}}_5)$ as well as $\mathsf{RMIN}(\widehat{\mathcal{H}}_8)$ are solvable in polynomial time.

6.5 Finding a Consistent Scenario

In order to decide the consistency of a set of RCC-8 constraints Θ, a path-consistent refinement of Θ containing only constraints over $\widehat{\mathcal{H}}_8$ must be found[8] (usually by applying backtracking). However, for other tasks such as finding a model of Θ this is not sufficient. For these tasks it is rather necessary to have base relations between any pair of variables involved in Θ, i.e., a consistent scenario for Θ is required.

A naive algorithm for finding a consistent scenario for Θ is based on iteratively selecting a label and enforcing path-consistency on the restricted set. This requires $O(n^5)$ time if Θ contains only constraints over $\widehat{\mathcal{H}}_8$, and to the best of our knowledge no algorithm with a better worst-case complexity is known. As in the case of qualitative temporal reasoning (e.g., [115, 155]),

---

[8] Of course also any other tractable subclass of RCC-8 for which path-consistency is sufficient for deciding consistency could be used.

by exploiting some properties of the particular class of used relations, it is possible to design more efficient algorithms. For instance, a more efficient algorithm is possible for a certain set of relations when path-consistency implies minimal labels or strong consistency [38] for this set. In this case any selected label guarantees consistency of the refined set. Therefore it is not necessary to choose another label of the same constraint and, thus, the algorithm requires only time $O(n^3)$. However, as it follows from Proposition 6.55, this is not the case for $\widehat{\mathcal{H}}_8$.

In the following we will prove that given a consistent set Θ of constraints over $\widehat{\mathcal{H}}_8$, a consistent scenario Θ_s can be obtained in $O(n^3)$ time by reducing all the constraints of Θ to constraints over the base relations in a very particular way. In order to prove this, we will use several *refinement strategies* of the form "given a particular set $\mathcal{S} \subseteq \widehat{\mathcal{H}}_8$, a relation R', and a path-consistent set Θ of constraints over $\mathcal{S}$, Θ can be consistently refined to Θ' by replacing each constraint xRy in Θ, such that $R' \subset R$, with the constraint $xR'y$." We will use refinement strategies for the following sets of relations. These sets are denoted $\widehat{\mathcal{H}}_{\downarrow_{R'}}$, in order to indicate that there is a refinement strategy using the relation R'.

Definition 6.56.

- $\widehat{\mathcal{H}}_{\downarrow \mathsf{DC}} = \widehat{\mathcal{H}}_8$
- $\widehat{\mathcal{H}}_{\downarrow \mathsf{EC}} = \{R \in \widehat{\mathcal{H}}_8 \mid R \text{ does not contain both } \mathsf{DC} \text{ and } \mathsf{EC}\}$
- $\widehat{\mathcal{H}}_{\downarrow \mathsf{PO}} = \{R \in \widehat{\mathcal{H}}_8 \mid R \text{ does not contain any of } \mathsf{DC}, \mathsf{EC}, \text{ or } \mathsf{EQ}, \\ \qquad\qquad\qquad \text{unless } R \text{ is a base relation}\}$
- $\widehat{\mathcal{H}}_{\downarrow \mathsf{NTPP}} = \{R \in \widehat{\mathcal{H}}_8 \mid R \text{ does not contain any of } \mathsf{DC}, \mathsf{EC}, \mathsf{EQ}, \text{ or } \mathsf{PO}, \\ \qquad\qquad\qquad\quad \text{unless } R \text{ is a base relation}\}$

By using the encoding of RCC-8 in classical propositional logic as specified in Proposition 6.19, it is possible to prove that these refinement strategies preserve consistency. For these proofs we need a property of R_{γ^*}–chains as defined in Definition 6.37. As a reminder, note that $c(\Theta)$ specifies the set of clauses corresponding to Θ and $p(m(\Theta))$ specifies the propositional encoding of the modal encoding $m(\Theta)$ of Θ.

Lemma 6.57. *Let Θ be a set of constraints over $\widehat{\mathcal{H}}_8$ that contains an R_{γ^*}– chain from x to y for some variables x and y. If for some w the clause $\{X_w^*\}$ can be derived from $c(\Theta)$ and $\{Y_w^*\}$ by using PUR, then $\{X_k^*\}$ can be derived from $c(\Theta)$ and $\{Y_k^*\}$ by using PUR for all k.*

Proof. Suppose that $\{X_w^*\}$ is derivable from $c(\Theta)$ and that zRv is a constraint of the R_{γ^*}–chain from x to y. Then either $\{Z_w^*, \neg V_w^*\}$ or $\{Z_w^*, \neg V_w^*, \neg Z_j^*, \neg V_j^*\}$ for some j is contained in $c(\Theta)$ and $\{Z_w^*\}$ must also be derivable from $c(\Theta)$ (otherwise $\{X_w^*\}$ would not be derivable). In order to derive $\{Z_w^*\}$, the positive unit clause $\{V_w^*\}$, or the positive unit clauses $\{V_w^*\}$, $\{Z_j^*\}$, and $\{V_j^*\}$ must be present, respectively. Since $\{Z_k^*, \neg V_k^*\}$ or $\{Z_k^*, \neg V_k^*, \neg Z_j^*, \neg V_j^*\}$, respectively, are also contained in $c(\Theta)$ for any k, $\{Z_k^*\}$ can be derived if $\{V_k^*\}$ is present.

Thus, if $\{Y_k^*\}$ is present, $\{X_k^*\}$ can be derived by propagating positive unit clauses involving k through the R_{γ^*}–chain. ∎

We are now prepared to prove refinement strategies for the different sets defined in Definition 6.56.

Lemma 6.58 (DC-refinement). *Let Θ be a path-consistent set of constraints over $\widehat{\mathcal{H}}_{\downarrow DC}$. Θ can be consistently refined to Θ' by replacing every constraint $xRy \in \Theta$ such that $\{DC\} \subset R$ with the constraint $x\{DC\}y$.*

Proof. Let xRy with $R = \{DC\} \cup R'$ be one of the constraints of Θ, and suppose that Θ becomes inconsistent if xRy is replaced with $x\{DC\}y$ resulting in Θ''. Since the propositional encoding of $x\{DC\}y$ is an indefinite Horn formula $(c(x\{DC\}y) = \bigcup_{w \in W_0}\{\{\neg X_w^*, \neg Y_w^*\}\})$, no new positive unit clause can be derived by using these Horn clauses. Thus, the empty clause can be derived from $c(\Theta'')$ by using PUR only when for some w both the unit clauses $\{X_w^*\}$ and $\{Y_w^*\}$ can be derived from $c(\Theta)$. It follows from Lemma 6.47 and Lemma 6.52 that if this were possible, then Θ would not be path-consistent since $\{DC\}$ would have been removed from xRy by applying a path-consistency algorithm. This contradicts our assumptions, so Θ'' is also consistent. Since no new positive unit clause is derivable from $c(\Theta'')$, any constraint xRy of Θ that contains $\{DC\}$ can be replaced with $x\{DC\}y$ simultaneously, for any pair of variables x and y in Θ, without applying a path-consistency algorithm after each refinement. ∎

The proofs of the following three refinement strategies are more complex than the proof of the previous one since there are more cases to consider. However, all cases can be handled with similar methods as used in the previous proof, namely, by looking at whether the changes to the propositional encodings resulting from the refinement of constraints permit to derive the empty clause by using PUR. In all of these cases it turns out (mostly by applying Lemma 6.57) that if the empty clause is derivable after the refinements, then it was also derivable before the refinement. Therefore the refinements preserve consistency of the set of constraints.

Lemma 6.59 (EC-refinement). *Let Θ be a path-consistent set of constraints over $\widehat{\mathcal{H}}_{\downarrow EC}$. Θ can be consistently refined to Θ' by replacing every constraint $xRy \in \Theta$ such that $\{EC\} \subset R$ with the constraint $x\{EC\}y$.*

Proof. Let xRy with $R = \{EC\} \cup R'$ and $\{DC\} \not\subset R'$ (this is guaranteed by the definition of $\widehat{\mathcal{H}}_{\downarrow EC}$) be one of the constraints of Θ and suppose that Θ becomes inconsistent if xRy is refined to $x\{EC\}y$ resulting in Θ''. $c(x\{EC\}y) = \{\{X_w\}, \{Y_w\}\} \cup \bigcup_{v \in W_0} \bigcup_{i=1}^{2n}\{\{\neg X_v^i, \neg Y_v^i\}\}$. Since R does not contain $\{DC\}$, x and y must have a common point and, therefore, $c(xRy)$ must contain the unit clauses $\{X_{w'}^*\}$ and $\{Y_{w'}^*\}$ for some w'. There are three possibilities of why Θ'' can be inconsistent:

1. $\{X_v^i\}$ and $\{Y_v^i\}$ for some v and some i can be derived from $c(\Theta'')$: The binary clauses of $c(x\{EC\}y)$ cannot be used to derive positive unit clauses. Thus, if $\{X_v^i\}$ and $\{Y_v^i\}$ can be derived from $c(\Theta'')$, they must be derived using $\{X_w\}$ and $\{Y_w\}$. By Lemma 6.57 they can also be derived using $\{X_{w'}^*\}$ and $\{Y_{w'}^*\}$, i.e., from $c(\Theta)$. As it was shown in the proof of Lemma 6.47, if $\{X_v^i\}$ and $\{Y_v^i\}$ can be derived from $c(\Theta)$, then a path-consistency algorithm eliminates $\{EC\}$ from xRy and, therefore Θ cannot be path-consistent, which is a contradiction.

2. $\{Z_w^*\}$ and $\{\neg Z_w^*\}$ for some spatial variable z are derived using $\{X_w\}$ and $\{Y_w\}$: By Lemma 6.57 they can also be derived using $\{X_{w'}^*\}$ and $\{Y_{w'}^*\}$, i.e., from $c(\Theta)$. Therefore Θ cannot be consistent, which is a contradiction.

3. $\{V_s^j\}$ and $\{\neg V_s^j\}$ for some s and some j can be derived from $c(\Theta'')$: If these clauses are derived using $\{X_w\}$ and $\{Y_w\}$, by Lemma 6.57 they can also be derived using $\{X_{w'}^*\}$ and $\{Y_{w'}^*\}$, i.e., from $c(\Theta)$. So, they must be derived using the binary clauses introduced by $c(x\{EC\}y)$, and, because we are using positive unit resolution, either $v \equiv x$, or $v \equiv y$. Suppose without loss of generality that $v \equiv x$. Then, in order to derive the clauses $\{X_s^j\}$ and $\{\neg X_s^j\}$, the clauses $\{X_s^j\}$ and $\{Y_s^j\}$ are required, which is equivalent to case 1.

So in all cases, Θ'' cannot be inconsistent if Θ is path-consistent and Θ does not contain any relation R with $\{DC, EC\} \subseteq R$. The reduction of R to $\{EC\}$ has no influence on the other relations containing $\{EC\}$, since no new positive unit clauses are derivable from $c(\Theta'')$ which are not already derivable from $c(\Theta)$. So this reduction can be done simultaneously for all relations containing $\{EC\}$.   ∎

Lemma 6.60 (PO-refinement). *Let Θ be a path-consistent set of constraints over $\widehat{\mathcal{H}}_{\downarrow PO}$. Θ can be consistently refined to Θ' by replacing every constraint $xRy \in \Theta$ such that $\{PO\} \subset R$ with the constraint $x\{PO\}y$.*

Proof. Let xRy with $R = \{PO\} \cup R'$, $R \in \widehat{\mathcal{H}}_{\downarrow PO}$ be one of the constraints of Θ and suppose that Θ becomes inconsistent if xRy is refined to $x\{PO\}y$ resulting in Θ''. We will prove by contradiction that Θ'' cannot be inconsistent if Θ is path-consistent using a case-analysis over the different possibilities of $R \in \widehat{\mathcal{H}}_{\downarrow PO}$. The relations of $\widehat{\mathcal{H}}_{\downarrow PO}$ are constructed as follows:

(a) The relations $\{PO, TPP\}$, $\{PO, TPP^{-1}\}$, $\{PO, TPP^{-1}, NTPP^{-1}\}$, and $\{PO, TPP, NTPP\}$ are contained in $\mathcal{H}_8$. Their abbreviated form is the following (see Appendix A.1):

$$m(\{PO, TPP\}) = \neg\gamma \wedge \neg\delta i \wedge \neg\pi i$$
$$m(\{PO, TPP, NTPP\}) = \neg\gamma \wedge \neg\delta i$$
$$m(\{PO, TPP^{-1}\}) = \neg\pi \wedge \neg\delta i \wedge \neg\gamma i$$
$$m(\{PO, TPP^{-1}, NTPP^{-1}\}) = \neg\pi \wedge \neg\delta i$$

(b) Many other relations of $\widehat{\mathcal{H}}_{\downarrow PO}$ are constructed as a conjunction of other relations such that one of the conjuncts is either one of the four relations of (a), or the relation $\{\overline{DC, EC}\}$ (see Appendix A.2):

$$\{PO, NTPP\} = \{PO, TPP, NTPP\} \cap \{\overline{TPP, EQ}\}$$
$$\{PO, NTPP^{-1}\} = \{PO, TPP^{-1}, NTPP^{-1}\} \cap \{\overline{TPP^{-1}, EQ}\}$$
$$\{PO, TPP, NTPP^{-1}\} = \{\overline{TPP^{-1}, EQ}\} \cap \{\overline{NTPP}\} \cap \{\overline{DC, EC}\}$$
$$\{PO, TPP, NTPP, NTPP^{-1}\} = \{\overline{TPP^{-1}, EQ}\} \cap \{\overline{DC, EC}\}$$
$$\{PO, NTPP, NTPP^{-1}\} = \{\overline{TPP, TPP^{-1}, EQ}\} \cap \{\overline{DC, EC}\}$$
$$\{PO, TPP, TPP^{-1}, NTPP^{-1}\} = \{\overline{NTPP, EQ}\} \cap \{\overline{DC, EC}\}$$
$$\{PO, NTPP, TPP^{-1}, NTPP^{-1}\} = \{\overline{TPP, EQ}\} \cap \{\overline{DC, EC}\}$$

(c) The relations $\{PO, TPP, TPP^{-1}\}$ and $\{PO, TPP, NTPP, TPP^{-1}\}$ are constructed by composition and conjunction of other relations as follows (see Appendix A.2):

$$\{PO, TPP, TPP^{-1}\} = (\{EC, PO, TPP^{-1}\} \circ \{TPP\}) \cap \{\overline{EQ}\}$$
$$\cap \{\overline{EC}\} \cap \{\overline{NTPP, NTPP^{-1}}\}$$
$$\{PO, TPP, NTPP, TPP^{-1}\} = (\{PO, TPP^{-1}\} \circ \{TPP\}) \cap \{\overline{EQ}\}$$

We will now prove by contradiction that xRy for all relations R of $\widehat{\mathcal{H}}_{\downarrow PO}$ can be refined to $x\{PO\}y$ without changing consistency of Θ ($m(PO) = \neg\pi \wedge \neg\gamma \wedge \neg\delta i$). We will start with the four relations of (a).

1. $R = \{PO, TPP\}$: Θ'' can only become inconsistent because $c(\neg\pi i) = \{\{X_w\}, \{\neg Y_w^{h(\neg y)}\}\}$ is replaced with $c(\neg\pi) = \{\{X_w\}, \{\neg Y_w\}\}$ in $c(\Theta'')$. In order to derive the empty clause by using PUR, $\{Y_w\}$ must be derivable from $c(\Theta'')$. This is only possible if there is an R_{γ^*}-chain from x to y. Then, $\{Y_w\}$ is also derivable from $c(\Theta)$. It follows from Lemma 6.47 that in this case a path-consistency algorithm eliminates $\{PO\}$ from R. Thus, Θ is not path-consistent, which is a contradiction.

2. $R = \{PO, TPP, NTPP\}$: Θ'' can only become inconsistent because $c(\neg\pi) = \{\{X_w\}, \{\neg Y_w\}\}$ is added to $c(\Theta)$ resulting in $c(\Theta'')$. With PUR the empty clause can only be derived from $c(\Theta'')$ if $\{Y_w\}$ is derivable, i.e., if there is an R_{γ^*}-chain from x to y. Then, $\{Y_w\}$ is also derivable from $c(\Theta)$ and it follows from Lemma 6.47 that in this case a path-consistency algorithm eliminates $\{PO\}$ from R. Thus, Θ is not path-consistent, which is a contradiction.

For the two converse relations $\{PO, TPP^{-1}\}$ and $\{PO, TPP^{-1}, NTPP^{-1}\}$ the proofs are analogous. The relations of (b) are conjunctions of either a relation of (a) or the relation $\{\overline{DC, EC}\}$ with other relations, i.e., $R = R' \cap R''$ where

R' is a relation of (a) or the relation $\{\overline{\mathsf{DC}, \mathsf{EC}}\}$. For all relations of (b), $c(\Theta'')$ emerges from $c(\Theta)$ by replacing $c(R')$ with $c(\{\mathsf{PO}\})$. Thus, for all relations of (b) where R' is a relation of (a) it follows from 1. and 2. that Θ'' is consistent if Θ is path-consistent. We will now prove that $c(\{\overline{\mathsf{DC}, \mathsf{EC}}\})$ can always be replaced with $c(\{\mathsf{PO}\})$ without making $c(\Theta'')$ unsatisfiable. Then, it follows that Θ'' is consistent if R is a relations of (b) such that $R' = \{\overline{\mathsf{DC}, \mathsf{EC}}\}$.

3. R is constructed using $\{\overline{\mathsf{DC}, \mathsf{EC}}\}$: $m(\overline{\{\mathsf{DC}, \mathsf{EC}\}}) = \neg\delta\mathrm{i}$, so Θ'' becomes inconsistent because $c(\neg\pi \wedge \neg\gamma)$ is added to $c(\Theta)$. If $c(\neg\pi)$ is used to derive the empty clause, the proof is the same as in case 2. If $c(\neg\gamma)$ is used to derive the empty clause, the proof is analogous to case 2, since $\neg\gamma$ is the converse of $\neg\pi$. Thus, if Θ'' is inconsistent, Θ cannot be path-consistent which is a contradiction.

The relations of (c) are intersections of either $\{\mathsf{EC}, \mathsf{PO}, \mathsf{TPP}^{-1}\} \circ \{\mathsf{TPP}\}$ or $\{\mathsf{PO}, \mathsf{TPP}^{-1}\} \circ \{\mathsf{TPP}\}$ with other relations, i.e., there is a pseudo variable z which is only related with x and y. We already know that $\{\mathsf{PO}, \mathsf{TPP}^{-1}\}$ can be refined to $\{\mathsf{PO}\}$ without changing consistency of Θ (see above), thus, $x\{\mathsf{PO}, \mathsf{TPP}^{-1}\} \circ \{\mathsf{TPP}\}y$ is refined to $x\{\mathsf{PO}\} \circ \{\mathsf{TPP}\}y = x\{\mathsf{PO}, \mathsf{TPP}, \mathsf{NTPP}\}y$, which is equivalent to case 2. If it can be shown that $x\{\mathsf{EC}, \mathsf{PO}, \mathsf{TPP}^{-1}\}z$ can be refined to $x\{\mathsf{PO}\}z$ without changing consistency of Θ (provided that z is a pseudo variable which is only related with x and y), then $x\{\mathsf{EC}, \mathsf{PO}, \mathsf{TPP}^{-1}\} \circ \{\mathsf{TPP}\}y$ is refined to $x\{\mathsf{PO}\} \circ \{\mathsf{TPP}\}y = x\{\mathsf{PO}, \mathsf{TPP}, \mathsf{NTPP}\}y$ and, thus, xRy is refined to $x\{\mathsf{PO}, \mathsf{TPP}\}y$, which is equivalent to case 1.

4. xRy is constructed using $x\{\mathsf{EC}, \mathsf{PO}, \mathsf{TPP}^{-1}\}z$ where z is a pseudo variable: $m(\{\mathsf{EC}, \mathsf{PO}, \mathsf{TPP}^{-1}\}) = \neg\delta \wedge \neg\pi \wedge \neg\gamma\mathrm{i}$, so Θ'' becomes inconsistent because $c(\neg\delta)$ is replaced with $c(\neg\delta\mathrm{i})$ or because $c(\neg\gamma\mathrm{i})$ is replaced with $c(\neg\gamma)$. If the empty clause is derivable by using $c(\neg\gamma)$, then the proof is analogous to case 1. If the empty clause is derivable by using $c(\neg\delta\mathrm{i}) = \bigcup_{i=1}^{2n}\{\{X_w^i\}, \{z_w^i\}\}$, then, since z is a pseudo variable, the empty clause must be derived by using $\{X_w^i\}$ and $\{Y_w^i\}$ for some i. This is possible only if there is a spatial variable u and a spatial variable v such that the relation between u and v is a subset of $\{\mathsf{DC}, \mathsf{EC}\}$ and if Θ contains an R_{γ^*}-chain from u to x and an R_{γ^*}-chain from v to y. In this case a path-consistency algorithm refines xRy to $x\{\mathsf{DC}, \mathsf{EC}\}y$, i.e., Θ is inconsistent which is a contradiction.

In all cases no new R_{γ^*}-chains are introduced, so by Lemma 6.57 no new positive unit clauses are derivable and, therefore, all relations of $\widehat{\mathcal{H}}_{\downarrow\mathsf{PO}}$ can be refined to $\{\mathsf{PO}\}$ simultaneously without changing consistency of Θ. ∎

Lemma 6.61 (NTPP-refinement). *Let Θ be a path-consistent set of constraints over $\widehat{\mathcal{H}}_{\downarrow\mathsf{NTPP}}$. Θ can be consistently refined to Θ' by replacing every constraint $xRy \in \Theta$ such that $\{\mathsf{NTPP}\} \subset R$ with the constraint $x\{\mathsf{NTPP}\}y$.*

Proof. Let $x\{\mathsf{TPP}^{-1}, \mathsf{NTPP}^{-1}\}y$ be one of the constraints of Θ and suppose that Θ becomes inconsistent if it is refined to $x\{\mathsf{NTPP}^{-1}\}y$ resulting in Θ''. The abbreviations of the two constraints are $m(\{\mathsf{TPP}^{-1}, \mathsf{NTPP}^{-1}\}) = \neg\pi \wedge \gamma$ and $m(\{\mathsf{NTPP}^{-1}\}) = \neg\pi \wedge \gamma\mathrm{i}$, so Θ'' becomes inconsistent because $c(\gamma) = \bigcup_{w \in W_0}\{\{X_w, \neg Y_w\}\} \cup \bigcup_{w \in W_0}\bigcup_{i=1}^{2n}\{\{X_w^i, \neg Y_w^i\}\}$ is replaced with $c(\gamma\mathrm{i}) = \bigcup_{w \in W_0}\bigcup_{i=1}^{2n}\{\{X_w^i, \neg Y_w\}\}$ in $c(\Theta'')$. Because of the closedness property $c(\mathsf{CP}) = \bigcup_{w \in W_0}\bigcup_{i=1}^{2n}\{\{X_w, \neg X_w^i\}\}$, the clause $\{X_w\}$ can be derived from $\{Y_w\}$ and $c(\gamma)$ as well as from $\{Y_w\}$ and $c(\gamma\mathrm{i})$, so the only clause that can be derived from $c(\Theta'')$ but not from $c(\Theta)$ is the clause $\{X_w^i\}$, provided that $\{Y_w^i\}$ cannot be derived from $c(\Theta)$.

We must find a way of how $\{X_w^i\}$ can be used to derive the empty clause such that (1) $\{X_w\}$ cannot be used to derive the empty clause, and such that (2) $\{Y_w^i\}$ cannot be derived. For this there must be four spatial variables u, u', v, v' with an $R_{\gamma*}$-chain from u to u' and an $R_{\gamma*}$-chain from v to v' such that xRy is part of the $R_{\gamma*}$-chain from v to v'. Because of (1), the relation between u and v must be one of $R_{\delta\mathrm{i}}$, but not of R_δ, i.e., $u\{\mathsf{EC}\}v$ must hold. Because of (2), the relation between u' and v' must be one of $R_{\neg\delta}$, but not of $R_{\neg\delta\mathrm{i}}$, i.e., $u'\{\mathsf{EC}\}v'$ must hold. Since Θ is path-consistent, the relation between u and u' and the relation between v and v' is one of R_γ. By composition, the relation between u' and v is a subset of $\{\mathsf{DC}, \mathsf{EC}\}$ and the relation between v and v' must be $\{\mathsf{TPP}\}$. Therefore, R must also be $\{\mathsf{TPP}\}$, thus, Θ is not path-consistent, which is a contradiction. The proof for the relation $\{\mathsf{TPP}, \mathsf{NTPP}\}$ is analogous.

Since Θ contains only constraints over $\widehat{\mathcal{H}}_{\downarrow\mathsf{NTPP}}$ and refining a constraint $x\{\mathsf{TPP}, \mathsf{NTPP}\}y$ to $x\{\mathsf{NTPP}\}y$ never makes a label NTPP of another constraint $u\{\mathsf{TPP}, \mathsf{NTPP}\}v$ inconsistent, all refinements can be made simultaneously. ∎

In addition to the four refinement strategies described above, we need a further constraint refinement technique for handling relations containing $\{\mathsf{EQ}\}$.

Lemma 6.62 (EQ-elimination). *Let Θ be a path-consistent set of constraints over $\widehat{\mathcal{H}}_8$. Θ can be consistently refined to Θ' by eliminating $\{\mathsf{EQ}\}$ from every constraint $xRy \in \Theta$ unless $R = \{\mathsf{EQ}\}$.*

Proof. Let xRy be one of the constraints of Θ, and suppose that Θ becomes inconsistent if $\{\mathsf{EQ}\}$ is eliminated from xRy resulting in Θ''. Since eliminating $\{\mathsf{EQ}\}$ from a relation R is equivalent to intersecting R with the relation $\overline{\{\mathsf{EQ}\}}$ expressible as $\overline{\{\mathsf{EQ}\}} = \{\mathsf{EC}\} \circ \{\mathsf{DC}, \mathsf{PO}\}$, Θ'' is equivalent to $\Theta \cup \{x\{\mathsf{EC}\}z, z\{\mathsf{DC}, \mathsf{PO}\}y\}$ where z is a pseudo variable which is only related with x and y. As neither $\{\mathsf{EC}\}$ nor $\{\mathsf{DC}, \mathsf{PO}\}$ are contained in $R_{\gamma*}$, no positive unit clauses for literals of $c(\Theta)$ can be derived from $c(\Theta'')$ by PUR using the clauses resulting from the propositional encoding of the two new constraints. Thus, the empty clause can only be derived from $c(\Theta'')$ by using the unit clauses in the propositional encodings of the new constraints. Because of Lemma 6.57 there must be a way in which only the unit clauses of the newly added constraints can be used to derive the empty clause, and not the unit clauses derivable from $c(\Theta)$. This is possible only if Θ contains an R_{γ^*}–chain from x to y and an R_{γ^*}–chain from y to x. But if this were the case, Θ would not be path-consistent, and thus the initial assumptions would be contradicted. Since no new R_{γ^*}–chain is introduced in Θ'', {EQ} can be eliminated from all relations simultaneously without applying the path-consistency algorithm after each elimination.   ∎

We can now combine the five strategies to derive an algorithm for determining a consistent scenario for a consistent set of constraints over $\widehat{\mathcal{H}}_8$.

Theorem 6.63. *For each path-consistent set Θ of constraints over $\widehat{\mathcal{H}}_8$, a consistent scenario Θ_s can be determined in time $O(n^3)$, where n is the number of variables involved in Θ.*

Proof. The following algorithm, SCENARIO(Θ), solves the problem:

(1) apply DC-refinement, (2) apply EC-refinement, (3) apply EQ-elimination, (4) apply PO-refinement, (5) apply NTPP-refinement, (6) return the set of the resulting constraints.
Impose path-consistency after each of the steps (1)–(5).

SCENARIO(Θ) terminates in $O(n^3)$ time since each of the steps (1)–(5) takes time $O(n^2)$ and path-consistency, which takes time $O(n^3)$, is computed 5 times. By Definition 6.56 we have that $\widehat{\mathcal{H}}_{\downarrow\text{NTPP}} \subset \widehat{\mathcal{H}}_{\downarrow\text{PO}} \subset \widehat{\mathcal{H}}_{\downarrow\text{EC}} \subset \widehat{\mathcal{H}}_8$. It follows from Lemmas 6.58 to 6.62 that after applying each refinement step, Θ is refined to a set of constraints containing only relations for which the next refinement step is guaranteed to make a consistent refinement. Since the interleaved applications of the path-consistency algorithm can only refine constraints and never add new base relations to the constraints, the possible set of relations obtained after each step is not extended by applying the path-consistency algorithm.

Thus, since Θ is consistent and contains only constraints over $\widehat{\mathcal{H}}_8$, the output of SCENARIO(Θ) is consistent. Moreover, since any (non-base) relation of $\widehat{\mathcal{H}}_8$ contains one of DC, EC, PO, NTPP, or NTPP^{-1} (see the definition of $\widehat{\mathcal{H}}_8$ in Section 6.3), the interleaved applications of path-consistency and steps (1)–(5) guarantee that the output of SCENARIO(Θ) is a consistent scenario for Θ.   ∎

By applying SCENARIO(Θ) to a path-consistent set of constraints over $\widehat{\mathcal{H}}_8$ we obtain a particular consistent scenario Θ_s of Θ. The properties of Θ_s are described in the following lemma.

Lemma 6.64. *Let Θ be a path-consistent set of constraints over $\widehat{\mathcal{H}}_8$ involving the variables x and y, and let Θ_s be the output of SCENARIO(Θ).*

- *The constraint $x\{\mathsf{EQ}\}y$ is contained in Θ_s only if it is also contained in Θ.*
- *The constraint $x\{\mathsf{TPP}\}y$ is contained in Θ_s only if Θ contains either $x\{\mathsf{TPP}\}y$ or $x\{\mathsf{TPP},\mathsf{EQ}\}y$.*
- *The constraint $x\{\mathsf{NTPP}\}y$ is contained in Θ_s only if Θ contains either $x\{\mathsf{NTPP}\}y$, $x\{\mathsf{NTPP},\mathsf{TPP}\}y$ or $x\{\mathsf{NTPP},\mathsf{TPP},\mathsf{EQ}\}y$.*

In all the other cases, $xRy \in \Theta$ is refined to one of $x\{\mathsf{DC}\}y$, $x\{\mathsf{EC}\}y$ or $x\{\mathsf{PO}\}y$ in Θ_s.

Proof Sketch. If the path-consistency algorithm were not applied after each of the steps (1) - (5) in SCENARIO(Θ), the proof would immediately follow from the applications of the refinements in the algorithm. Considering the interleaved path-consistency computations, it might be possible that after refining a constraint of Θ to one of $\{\mathsf{DC}\}, \{\mathsf{EC}\}$, or $\{\mathsf{PO}\}$ by one of the steps (1), (2), or (4), another constraint $xRy \in \Theta$, such that $R \cap \{\mathsf{DC}, \mathsf{EC}, \mathsf{PO}\} \neq \emptyset$ and $R \cap \{\mathsf{TPP}, \mathsf{TPP}^{-1}, \mathsf{NTPP}, \mathsf{NTPP}^{-1}, \mathsf{EQ}\} \neq \emptyset$, is refined by the path-consistency algorithm to $xR'y$ such that $R' \cap \{\mathsf{DC}, \mathsf{EC}, \mathsf{PO}\} = \emptyset$. In this case, xRy would be refined by SCENARIO(Θ) to one of $x\{\mathsf{NTPP}\}y$, $x\{\mathsf{TPP}\}y$, $x\{\mathsf{NTPP}^{-1}\}y$, $x\{\mathsf{TPP}^{-1}\}y$, or $x\{\mathsf{EQ}\}y$ in Θ_s which contradicts the lemma. However, by analyzing the composition table of the RCC-8 relations (see Table 4.2) it follows that this is never possible for the sets of relations used by the different refinement strategies. ∎

6.6 Discussion

In this chapter we first showed that reasoning over RCC-8 constraints is NP-hard. By analyzing our modification of Bennett's encoding of RCC-8 into modal logic, we found that whenever a set of RCC-8 constraints is consistent, the corresponding modal formula has a Kripke model of a particular type, the RCC-8-model. Based on this model, we transformed the modal encoding of RCC-8 into a classical propositional formula (which demonstrates that RSAT is in NP). Applying this encoding of RCC-8 into a propositional formula, we found that constraints over some RCC-8 relations are transformable to propositional Horn formulas, thus, deciding consistency of a set of constraints over these relations, denoted $\widehat{\mathcal{H}}_8$, is tractable. We further showed that $\widehat{\mathcal{H}}_8$, which contains 148 different RCC-8 relations including all base relations, is a maximal tractable subset of RCC-8, i.e., whenever any other relation is added to the set, deciding consistency becomes NP-hard. Hence, $\widehat{\mathcal{H}}_8$ marks the boundary between tractable and intractable subsets of $\widehat{\mathcal{H}}_8$. By relating positive unit resolution on the propositional encoding of RCC-8 to computing path-consistency, it turned out that path-consistency is sufficient for deciding consistency of sets of $\widehat{\mathcal{H}}_8$ constraints. Using the same proof technique, we identified an $O(n^3)$ algorithm for computing a consistent scenario of these sets.

One obvious advantage of the maximal tractable subset $\widehat{\mathcal{H}}_8$ is that the path-consistency method is sufficient for deciding RSAT if it is possible to restrict the relations used in an application to the relations of $\widehat{\mathcal{H}}_8$. In many applications this is certainly not possible. In spatial configuration tasks or queries to spatial databases, e.g., RCC-8 base relations and negations of the base relations are often used, as in *Find a region which is in A but does not overlap B*. Since the closure of these 16 relations is the whole set of 256 RCC-8 relations, all relations have to be considered in this kind of applications.

As in the case of temporal reasoning where the usage of the maximal tractable subset ORD-HORN has been extensively studied [127], $\widehat{\mathcal{H}}_8$ can also be used to speed up backtracking algorithms for the general NP-complete RSAT problem (see Section 2.4.2). Previously, every spatial constraint had to be refined to a base relation before the path-consistency method could be applied to decide consistency. In the worst case this has to be done for all possible refinements. Supposing that the relations are uniformly distributed, the average branching factor, i.e., the average number of different refinements of a single relation to relations of $\mathcal{B}$ is 4.0.

Using our results it is sufficient to make refinements of all relations to relations of $\widehat{\mathcal{H}}_8$. Except for four relations, each of the 108 relations not contained in $\widehat{\mathcal{H}}_8$ can be expressed as a union of two relations of $\widehat{\mathcal{H}}_8$, the four relations can only be expressed as a union of three $\widehat{\mathcal{H}}_8$ relations. This reduces the average branching factor to 1.4375 ($=(148 + 104 \times 2 + 4 \times 3)/256$). Both branching factors are of course worst-case measures because the search space can be considerably reduced when path-consistency is used as a forward checking method [112]. Table 6.1 shows the average size of the search space for the average branching factors given above. The size of the search space is computed as $b^{(n^2-n)/2}$ where b is the average branching factor, n the number of spatial variables contained in Θ, and $(n^2 - n)/2$ the number of different constraints. In Chapter 8 we present an empirical study of reaso-

Table 6.1. Comparison of the average size of the search space

$|Var(\Theta)|$	$\mathcal{B}$ (4.0)	$\widehat{\mathcal{H}}_8$ (1.4375)
5	10^6	38
7	4.4×10^{12}	2041
10	1.2×10^{27}	1.2×10^7

ning with RCC-8 where we also discuss the question of whether using $\widehat{\mathcal{H}}_8$ for backtracking is as good as Table 6.1 suggests.

Other complexity results on qualitative spatial reasoning were obtained by Grigni et al. [77] who worked with Egenhofer's topological relations [44]. Grigni et al [77] considered two different notions of satisfiability, relational consistency and realizability (see Section 4.4), which are both different from what we call consistency. Both kinds of satisfiability were found to be NP-hard for Egenhofer's relations. These NP-hardness results are independent from our NP-hardness result. The notion of realizability is much more constraining than our notion of consistency. It is also computationally much harder – realizability for the eight base relations of RCC-8, e.g., is not known to be in NP. "Realizability", i.e., finding one-piece regions, in three and higher dimensional space, however, is equivalent to our notion of consistency (see Chapter 9).

Nebel [126] proved tractability for the set of RCC-8 base relations by transforming the propositional intuitionistic encoding of the base relations given by Bennett [10] to Krom formulas. This tractability result, however, is not applicable   in our case, since Nebel did not consider the regularity condition. The regularity condition is necessary to rule out certain counterintuitive regions, as, e.g, regions which only consist of a boundary. Moreover, since we are restricting our analysis to closed regions, the regularity condition is required in order to guarantee inferences according to the RCC-8 composition table. Consider, e.g., a non-regular region B with a spike consisting of a piece of the boundary, like a balloon with a cord, such that only the spike intersects the two regions A and C where $A\{\text{NTPP}\}C$. In this case, since B intersects A and C but the interior of B does not intersect A and C, B is externally connected to both A and C, which is not consistent with the composition table.

Jonsson and Drakengren [101] made a complete classification of tractability of RCC-5. Apart from $\widehat{\mathcal{H}}_5$, the only maximal tractable subclass of RCC-5 containing all base relations, they discovered three other tractable subclasses not containing all base relations. For all subclasses not contained in one of the four tractable subsets, RSAT is NP-complete. While a complete classification of tractability is certainly worthwhile from a theoretical point of view, the practical usage of subclasses not containing all base relations is limited, as it is not possible to express definite knowledge within a given calculus even if it is available. Furthermore, these subclasses cannot be used to speed up backtracking algorithms as it is not possible to refine every relation to relations of these subclasses. So far we have neither identified other maximal tractable subclasses containing all base relations nor proved that $\widehat{\mathcal{H}}_8$ is the only such set. However, in the following chapter we will make a complete analysis of tractability by identifying all maximal tractable subclasses of RCC-8 that contain all base relations.

7. A Complete Analysis of Tractability in RCC-8

In Chapter 6 we identified $\widehat{\mathcal{H}}_8$, a maximal tractable subset of RCC-8 that contains all base relations and for which path-consistency is complete for deciding inconsistency. It is, however, not clear whether $\widehat{\mathcal{H}}_8$ is the only such set or whether there are other maximal tractable subsets of RCC-8 that contain all base relations. There are at least two reasons why answering this question is important. First of all, without a complete analysis of tractability there are interesting subsets of RCC-8 for which it is unknown whether reasoning over these sets is tractable or NP-complete. Secondly, if new tractable subsets can be identified, these can be used to further enhance efficiency of reasoning over RCC-8.

In this chapter we will give a complete analysis of tractability in RCC-8. Since only tractable subsets containing all base relations can be used to speed up backtracking algorithms for the full set of RCC-8 relations, we restrict our analysis to those subsets that contain all base relations.

Tractability of reasoning over $\widehat{\mathcal{H}}_8$ and also over ORD-Horn, the only maximal tractable subset of Allen's interval algebra containing all base relations [128] (see Section 2.5), was shown by reducing the consistency problem to propositional satisfiability (SAT) and identifying those relations that reduce to Horn formulas. Sufficiency of path-consistency for reasoning over these sets was shown by relating path-consistency to positive unit resolution over the corresponding Horn formulas. This way of proving tractability requires very detailed semantical knowledge about the considered set of relations and a suitable reduction to a tractable decision problem.

Instead of trying to find a different reduction for every possibly tractable subset of RCC-8, we develop in this chapter a method for proving tractability and sufficiency of path-consistency for reasoning over sets of relations. This method works not only for RSAT but in general for CSPSAT($\mathcal{S}$) where $\mathcal{S} \subseteq 2^{\mathcal{A}}$ and $\mathcal{A}$ is a set of JEPD relations which are atoms of a relation algebra. The method is simple, since it does not require other semantical knowledge about the relations than given by the composition table. It is automatic in the sense that it is not necessary to find a suitable reduction by hand, rather it is possible to either generate or verify a reduction automatically. A prerequisite of this method is a known tractable set for which path-consistency decides consistency.

J. Renz: Qualitative Spatial Reasoning with Topological Information, LNAI 2293, pp. 117–130, 2002.
© Springer-Verlag Berlin Heidelberg 2002

7.1 A General Method for Proving Tractability of Sets of Relations

In this section we develop a general method for proving tractability of reasoning over disjunctions of a JEPD set $\mathcal{A}$ of binary relations over a domain $\mathcal{D}$ which are atoms of a relation algebra, i.e., a method for proving tractability of $\mathsf{CSPSAT}(\mathcal{S})$ for sets $\mathcal{S} \subseteq 2^{\mathcal{A}}$ (see Section 2.4). In order to do so, our method requires a subset $\mathcal{T}$ of $2^{\mathcal{A}}$ for which path-consistency is already known to decide $\mathsf{CSPSAT}(\mathcal{T})$. Then the method checks whether it is possible to refine every constraint involving a relation in $\mathcal{S}$ according to a particular refinement scheme to a constraint involving a relation in $\mathcal{T}$ without changing consistency. The following definition will be central for our method.

Definition 7.1 (reduction by refinement).
Let $\mathcal{S}, \mathcal{T} \subseteq 2^{\mathcal{A}}$. $\mathcal{S}$ can be reduced by refinement to $\mathcal{T}$, if the following two conditions are satisfied:

1. *for every relation $S \in \mathcal{S}$ there is a relation $T_S \in \mathcal{T}$ with $T_S \subseteq S$,*
2. *every path-consistent set Θ of constraints over $\mathcal{S}$ can be refined to a set Θ' of constraints over $\mathcal{T}$ by replacing $x_i S x_j \in \Theta$ with $x_i T_S x_j \in \Theta'$ for $i < j$,[1] such that enforcing path-consistency to Θ' does not result in an inconsistency.*

This property of a set of relations can be used to derive its tractability.

Lemma 7.2. *If path-consistency decides $\mathsf{CSPSAT}(\mathcal{T})$ for a set $\mathcal{T} \subseteq 2^{\mathcal{A}}$, and $\mathcal{S}$ can be reduced by refinement to $\mathcal{T}$, then path-consistency decides $\mathsf{CSPSAT}(\mathcal{S})$.*

Proof. Let Θ be a path-consistent set of constraints over $\mathcal{S}$. Since $\mathcal{S}$ can be reduced by refinement to $\mathcal{T}$, there is by definition a set Θ' of constraints over $\mathcal{T}$ which is a refinement of Θ such that enforcing path-consistency to Θ' does not result in an inconsistency. Path-consistency decides $\mathsf{CSPSAT}(\mathcal{T})$, so Θ' is consistent, and, hence, Θ is also consistent. ∎

Since path-consistency can be enforced in cubic time, it is sufficient for proving tractability of $\mathsf{CSPSAT}(\mathcal{S})$ to show that $\mathcal{S}$ can be reduced by refinement to a set $\mathcal{T}$ for which path-consistency decides $\mathsf{CSPSAT}(\mathcal{T})$. Note that for refining a constraint xSy ($S \in \mathcal{S}$) to a constraint xT_Sy ($T_S \in \mathcal{T}$), it is not required that T_S is also contained in $\mathcal{S}$. Thus, with respect to common relations the two sets $\mathcal{S}$ and $\mathcal{T}$ are independent of each other. This is in contrast to Theorem 6.5 which allows to inherit tractability of a set of relations to its closure.

---

[1] Note that we only refine constraints $x_i S x_j$ for $i < j$. This is no restriction, as by enforcing path-consistency the converse constraint $x_j S^\smile x_i$ will also be refined. Rather it offers the possibility of refining, e.g., converse relations to other than converse sub-relations, i.e., if, for instance, R is refined to r, we can refine $R^\smile$ to a relation other than $r^\smile$.

We will now develop a method for showing that a set of relations $S \subseteq 2^{\mathcal{A}}$ can be reduced by refinement to another set $\mathcal{T} \subseteq 2^{\mathcal{A}}$. In order to manage the different refinements, we introduce a *refinement matrix* that contains for every relation $S \in \mathcal{S}$ all specified refinements.

Definition 7.3 (refinement matrix).
A refinement matrix M of S has $|\mathcal{S}| \times 2^{|\mathcal{A}|}$ Boolean entries such that for $S \in \mathcal{S}, R \in 2^{\mathcal{A}}, M[S][R] = true$ only if $R \subseteq S$.

For example, if the relation $\{\mathsf{DC}, \mathsf{EC}, \mathsf{PO}, \mathsf{TPP}\}$ is allowed to be refined only to the relations $\{\mathsf{DC}, \mathsf{TPP}\}$ and $\{\mathsf{DC}\}$, then $M[\{\mathsf{DC}, \mathsf{EC}, \mathsf{PO}, \mathsf{TPP}\}][R]$ is *true* only for $R = \{\mathsf{DC}, \mathsf{TPP}\}$ and for $R = \{\mathsf{DC}\}$ and *false* for all other relations $R \in 2^{\mathcal{A}}$. M is called the *basic refinement matrix* if $M[S][R] = true$ if and only if $S = R$.

We propose the algorithm CHECK-REFINEMENTS (see Figure 7.1) which takes as input a set of relations $\mathcal{S}$ and a refinement matrix M of $\mathcal{S}$. This algorithm computes all possible path-consistent triples of relations R_{12}, R_{23}, R_{13} of $\mathcal{S}$ (step 4) and enforces path-consistency (using a standard procedure PATH-CONSISTENCY) to every refinement $R'_{12}, R'_{23}, R'_{13}$ for which $M[R_{ij}][R'_{ij}] = true$ for all $i, j \in \{1, 2, 3\}, i < j$ (steps 5,6). If one of these refinements results in the empty relation, the algorithm returns `fail` (step 7). Otherwise, the resulting relations $R''_{12}, R''_{23}, R''_{13}$ are added to M by setting $M[R_{ij}][R''_{ij}] = true$ for all $i, j \in \{1, 2, 3\}, i < j$ (step 8). This is repeated until M has reached a fixed point (step 9), i.e., enforcing path-consistency on any possible refinement does not result in new relations anymore. If no inconsistency is detected in this process, the algorithm returns `succeed`.

A similar algorithm, GET-REFINEMENTS, returns the revised refinement matrix if CHECK-REFINEMENTS returns `succeed` and the basic refinement matrix if CHECK-REFINEMENTS returns `fail`. Since $\mathcal{A}$ is a finite set of relations, M can be changed only a finite number of times, so both algorithms always terminate. If $n = |2^{\mathcal{A}}|$ is the total number of relations, then there are at most n^3 possible triples of relations in step 4, at most n^3 possible refinements of each triple in step 5, and at most n^2 iterations of the *while* loop. Thus, a rough estimation of the worst-case running time of both algorithms leads to $O(n^8)$.

Lemma 7.4. *Let Θ be a path-consistent set of constraints over $\mathcal{S}$ and M a refinement matrix of $\mathcal{S}$. For every refinement Θ' of Θ with $x_i R' x_j \in \Theta'$ only if $x_i R x_j \in \Theta$, $i < j$, and $M[R][R'] = true$: if CHECK-REFINEMENTS$(\mathcal{S}, M)$ returns `succeed`, enforcing path-consistency to Θ' does not result in an inconsistency.*

Proof. Suppose that CHECK-REFINEMENTS$(\mathcal{S}, M)$ returns `succeed` and GET-REFINEMENTS$(\mathcal{S}, M)$ returns the refinement matrix M'. Suppose further that enforcing path-consistency to Θ' results in an inconsistency which is detected first for the three variables x_a, x_b, x_c. Suppose that for every pair of variables x_i, x_j, $x_i R_{ij} x_j \in \Theta$ and $x_i R'_{ij} x_j \in \Theta'$. Enforcing path-consistency

---

Algorithm: CHECK-REFINEMENTS
Input: A set $\mathcal{S}$ and a refinement matrix M of $\mathcal{S}$.
Output: **fail** if the refinements specified in M can make a path-consistent triple
of constraints over $\mathcal{S}$ inconsistent; **succeed** otherwise.

1. changes $\leftarrow$ *true*
2. *while* changes *do*
3. $oldM \leftarrow M$
4.   *for every* path-consistent triple
       $T = (R_{12}, R_{23}, R_{13})$ of relations over $\mathcal{S}$ *do*
5.   *for every* refinement $T' = (R'_{12}, R'_{23}, R'_{13})$ of T
           with $oldM[R_{12}][R'_{12}] = oldM[R_{23}][R'_{23}] =$
           $oldM[R_{13}][R'_{13}] = true$ *do*
6. $T'' \leftarrow$ PATH-CONSISTENCY(T')
7.   *if* $T'' = (R''_{12}, R''_{23}, R''_{13})$ contains the empty
               relation *then return* **fail**
8.   *else do* $M[R_{12}][R''_{12}] \leftarrow true,$
                         $M[R_{23}][R''_{23}] \leftarrow true,$
                         $M[R_{13}][R''_{13}] \leftarrow true$
9.   *if* $M = oldM$ *then* changes $\leftarrow$ *false*
10. *return* **succeed**

Fig. 7.1. Algorithm CHECK-REFINEMENTS

---

to Θ' can be done by successively enforcing path-consistency to every triple of variables of Θ'. Suppose that we start with the triple x_1, x_2, x_3.
We have that $M'[R_{ij}][R'_{ij}] = true$ for every $i, j \in \{1, 2, 3\}, i < j$. After enforcing path-consistency to this triple we obtain the relations R''_{ij} for
which again $M'[R_{ij}][R''_{ij}] = true$, otherwise CHECK-REFINEMENTS$(\mathcal{S}, M)$
would have returned **fail**. The same holds when we proceed with enforcing
path-consistency to every triple of variables, we always end up with relations R^*_{ij} for which $M'[R_{ij}][R^*_{ij}] = true$. Therefore, for all possible relations
$R^*_{ab}, R^*_{bc}, R^*_{ac}$ we can obtain by enforcing path-consistency to Θ' we have that
$M'[R_{ij}][R^*_{ij}] = true$ for all $i, j \in \{a, b, c\}, i < j$. If this triple of relations were
inconsistent, CHECK-REFINEMENTS$(\mathcal{S}, M)$ would have returned **fail**.   ∎

If CHECK-REFINEMENTS returns **succeed** and GET-REFINEMENTS returns
M', we have pre-computed all possible refinements of every path-consistent
triple of variables as given in the refinement matrix M'. Thus, applying these
refinements to a path-consistent set of constraints can never result in an
inconsistency when enforcing path-consistency.

Theorem 7.5. *Let* $\mathcal{S}, \mathcal{T} \subseteq 2^{\mathcal{A}}$, *and let* M *be a refinement matrix of* $\mathcal{S}$. GET-
REFINEMENTS$(\mathcal{S}, M)$ *returns the refinement matrix* M'. *If for every* $S \in \mathcal{S}$
there is a $T_S \in \mathcal{T}$ *with* $M'[S][T_S] = true$, *then* $\mathcal{S}$ *can be reduced by refinement
to* $\mathcal{T}$.

Proof. By the given conditions, we can refine every path-consistent set Θ of constraints over $\mathcal{S}$ to a set Θ' of constraints over $\mathcal{T}$ such that $M'[R_{ij}][R'_{ij}] = $ *true* for every $x_i R_{ij} x_j \in \Theta$ and $x_i R'_{ij} x_j \in \Theta'$, $i < j$, $R'_{ij} \in \mathcal{T}$. M' is a fixed point with respect to GET-REFINEMENTS, i.e., GET-REFINEMENTS$(\mathcal{S}, M')$ returns M', thus it follows from Lemma 7.4 that enforcing path-consistency to Θ' does not result in an inconsistency.   ∎

By Lemma 7.2 and Theorem 7.5 we have that the procedures CHECK-RE-FINEMENTS and GET-REFINEMENTS can be used to prove tractability for sets of relations.

Corollary 7.6. *Let $\mathcal{S}, \mathcal{T} \subseteq 2^{\mathcal{A}}$ be two sets such that path-consistency decides* CSPSAT$(\mathcal{T})$, *and let M be a refinement matrix of $\mathcal{S}$.* GET-REFINEMENTS$(\mathcal{S},$ $M)$ *returns M'. If for every $S \in \mathcal{S}$ there is a $T_S \in \mathcal{T}$ with $M'[S][T_S] = true$, then path-consistency decides* CSPSAT$(\mathcal{S})$.

If a suitable refinement matrix can be found, CHECK-REFINEMENTS can be used to immediately verify that reasoning over the given set of relations is tractable. One problem with this method is that though polynomial the algorithms are not very efficient. Especially for large sets of relations the algorithms are very slow. Fortunately, the algorithms are used for determining tractability of reasoning over sets of relations and not for the reasoning process itself.

Nevertheless, we propose in the following the algorithm CHECK-REFINE-MENTS-2 (see Figure 7.2), a modified and more efficient (but also more limited) version of CHECK-REFINEMENTS which is not based on refinement matrices that can handle all possible refinements, but on a one-dimensional *refinement array* that carries only one of the possible refinements.

Definition 7.7 (refinement array).
A one-dimensional refinement array A *of $\mathcal{S}$ has $|\mathcal{S}|$ entries with $A : \mathcal{S} \mapsto 2^{\mathcal{A}}$ such that $A[S] = R$ only if $S \in \mathcal{S}$ and $R \subseteq S$.*

A is called the *basic refinement array* if and only if $A[S] = S$ for all $S \in \mathcal{S}$.

CHECK-REFINEMENTS-2 differs from CHECK-REFINEMENTS-2 in that it uses only one possible refinement of every path-consistent triple of relations as given in the refinement array. The algorithm is focused towards finding the most restricted refinement of every relation and the refinement array is modified accordingly. If a fixed point is reached, i.e., if the refinement array cannot be modified anymore and does not contain the empty relation, then CHECK-REFINEMENTS-2 returns `succeed`. A similar algorithm GET-REFINEMENTS-2 returns the revised refinement array if CHECK-REFINEMENTS-2 returns `succeed` and the basic refinement array if CHECK-REFINEMENTS-2 returns `fail`. Both algorithms are much faster than CHECK-REFINEMENTS and GET-REFINEMENTS. The refinement array can be modified at most $n \log n$ times and at most n^3 triples have to be checked, so the worst-case running time of both algorithms is $O(n^4 \log n)$.

Algorithm: CHECK-REFINEMENTS-2

Input: A set $\mathcal{S}$ and a refinement array A of $\mathcal{S}$.

Output: `fail` if the refinements specified in A can make a path-consistent triple of constraints over $\mathcal{S}$ inconsistent; `succeed` otherwise.

1. changes $\leftarrow$ *true*
2. *while* changes *do*
3. $oldA \leftarrow A$
4.   *for every* path-consistent triple
       $T = (R_{12}, R_{23}, R_{13})$ of relations over $\mathcal{S}$ *do*
5.   refine T to $T' = (R'_{12}, R'_{23}, R'_{13})$
           with $A[R_{12}] = R'_{12}, A[R_{23}] = R'_{23}, A[R_{13}] = R'_{13}$
6. $T'' \leftarrow$ PATH-CONSISTENCY(T')
7.   *if* $T'' = (R''_{12}, R''_{23}, R''_{13})$ contains the empty
           relation *then return* `fail`
8.   *else do* $A[R_{12}] \leftarrow R''_{12}$,
                      $A[R_{23}] \leftarrow R''_{23}$,
                      $A[R_{13}] \leftarrow R''_{13}$
9.   *if* $A = oldA$ *then* changes $\leftarrow$ *false*
10. *return* `succeed`

Fig. 7.2. Algorithm CHECK-REFINEMENTS-2

If the initial refinement array contains only relations of a set $\mathcal{T}$, then all relations of the modified refinement array are obtained by composition, intersection, or converse of relations of $\mathcal{T}$, i.e., the modified relations are contained in $\widehat{\mathcal{T}}$.

Lemma 7.8. *Let Θ be a path-consistent set of constraints over $\mathcal{S}$ and A be a refinement array of $\mathcal{S}$.* GET-REFINEMENTS-2$(\mathcal{S}, A)$ *returns A'. If Θ' is obtained from Θ by replacing every constraint $xSy \in \Theta$ with $xA'[S]y \in \Theta'$, then Θ' is also path-consistent.*

Proof. A' is a fixed point of GET-REFINEMENTS-2, i.e., GET-REFINEMENTS-2$(\mathcal{S}, A')$ returns A'. Suppose that Θ' is not path-consistent, i.e., there is a triple of relations, say $R'_{12}, R'_{23}, R'_{13}$, which is refined to $R''_{12}, R''_{23}, R''_{13}$ by enforcing path-consistency to Θ'. If A' is the basic refinement array, Θ' is equal to Θ and, therefore, path-consistent. Otherwise, GET-REFINEMENTS-2 would have made the same refinements, thus, A' cannot be a fixed point of GET-REFINEMENTS-2. ∎

From the conditions of Definition 7.1 we obtain the following theorem.

Theorem 7.9. *Let $\mathcal{S} \subseteq 2^{\mathcal{A}}$ and A be a refinement array of $\mathcal{S}$. If* GET-REFINEMENTS-2$(\mathcal{S}, A)$ *returns A' which contains only relations of $\mathcal{T} \subseteq 2^{\mathcal{A}}$, then $\mathcal{S}$ can be reduced by refinement to $\mathcal{T}$.*

From Theorem 7.9 and Lemma 7.2 it follows that CHECK-REFINEMENTS-2 and GET-REFINEMENTS-2 can also be used to prove tractability of reasoning over sets of relations.

Corollary 7.10. *Let $\mathcal{S}, \mathcal{T} \subseteq 2^{\mathcal{A}}$ be two sets such that path-consistency decides* $\mathsf{CSPSAT}(\mathcal{T})$, *and let A be a refinement array of $\mathcal{S}$.* GET-REFINEMENTS-$2(\mathcal{S}, A)$ *returns A'. If A' contains only relations of $\mathcal{T}$, then path-consistency decides* $\mathsf{CSPSAT}(\mathcal{S})$.

In the following sections we apply our method to the Region Connection Calculus RCC-8 and to Allen's interval algebra for proving certain subsets to be tractable. For RCC-8 it will help us to make a complete analysis of tractability by identifying two large new maximal tractable subsets. Later, it will turn out that our method can be used for identifying the fastest possible algorithms for finding a consistent scenario.

7.2 Candidates for Maximal Tractable Subsets of RCC-8

In order to identify maximal tractable subsets of a set of relations with an NP-hard consistency problem, two different tasks have to be done. On the one hand, tractable subsets have to be identified. On the other hand, NP-hard subsets have to be identified. Then, a tractable subset is maximal tractable if any superset is NP-hard. If tractability has not yet been shown for certain subsets, the NP-hardness results restrict the number of different subsets that still have to be analyzed for tractability.

In this section we present new NP-hardness results for RCC-8 and identify those subsets that are candidates for maximal tractable subsets, i.e., if any other relation is added to one of those candidates they become NP-hard.

In Chapter 6, in Lemma 6.29 we proved maximality of $\widehat{\mathcal{H}}_8$ by showing that $\mathsf{RSAT}(\widehat{\mathcal{H}}_8 \cup \{\mathsf{EQ}, \mathsf{NTPP}\})$ is NP-hard and that the closure of $\widehat{\mathcal{H}}_8$ plus any relation of $\mathcal{N}_3$ contains the relation $\{\mathsf{EQ}, \mathsf{NTPP}\}$. This NP-hardness proof, however, does not make use of all relations of $\widehat{\mathcal{H}}_8$, but only of the relations $\{\mathsf{EC}, \mathsf{TPP}\}$ and $\{\mathsf{EC}, \mathsf{NTPP}\}$. Thus, any set of RCC-8 relations that contains all base relations plus the relations $\{\mathsf{EC}, \mathsf{TPP}\}$, $\{\mathsf{EC}, \mathsf{NTPP}\}$, and $\{\mathsf{EQ}, \mathsf{NTPP}\}$ is NP-hard.

This result rules out a lot of subsets of RCC-8 from being tractable, but still leaves the problem of tractability open for a large number of subsets. We found, however, that it is not necessary that both relations, $\{\mathsf{EC}, \mathsf{TPP}\}$ and $\{\mathsf{EC}, \mathsf{NTPP}\}$, must be added to $\{\mathsf{EQ}, \mathsf{NTPP}\}$ in order to enforce NP-hardness. It is rather sufficient for NP-hardness that only one of them is added to $\{\mathsf{EQ}, \mathsf{NTPP}\}$ which further reduces the possible number of tractable subsets.

The additional NP-hardness results are shown in the following two lemmata. They are proven by reducing 3SAT to the respective consistency problem. Both reductions are similar to the reductions given in Chapter 6: every literal as well as every literal occurrence L is reduced to a constraint $\mathsf{X}_L R \mathsf{Y}_L$ where $R = R_t \cup R_f$ and $R_t \cap R_f = \emptyset$. L is true iff $\mathsf{X}_L R_t \mathsf{Y}_L$ holds and false iff $\mathsf{X}_L R_f \mathsf{Y}_L$ holds. "Polarity constraints" enforce that $\mathsf{X}_{\neg L} R_t \mathsf{Y}_{\neg L}$ holds iff $X_L R_f Y_L$ holds, and *vice versa*. "Clause constraints" enforce that at least one literal of every clause is true.

Lemma 7.11. *If* $(\mathcal{B} \cup \{\{EQ, NTPP\}, \{EC, NTPP\}\}) \subseteq \mathcal{S}$, *then* $\mathsf{RSAT}(\mathcal{S})$ *is NP-complete.*

Proof Sketch. Transformation of 3SAT to $\mathsf{RSAT}(\mathcal{S})$. $R_t = \{NTPP\}$ and $R_f = \{EQ\}$. Polarity constraints:

$$X_L\{EC\}X_{\neg L}, Y_L\overline{\{DC, EC\}}Y_{\neg L},$$
$$X_L\{EC, NTPP\}Y_{\neg L}, X_{\neg L}\{EC, NTPP\}Y_L.$$

Clause constraints for each clause $c = \{i, j, k\}$:

$$Y_i\{NTPP^{-1}\}X_j, Y_j\{NTPP^{-1}\}X_k, Y_k\{NTPP^{-1}\}X_i.$$

$\overline{\{DC, EC\}}$ is contained in $\widehat{\mathcal{S}}$. Thus, $\mathsf{RSAT}(\widehat{\mathcal{S}})$ is NP-complete and because of Corollary 6.6, $\mathsf{RSAT}(\mathcal{S})$ is also NP-complete. ∎

Lemma 7.12. *If* $(\mathcal{B} \cup \{\{EQ, NTPP\}, \{EC, TPP\}\}) \subseteq \mathcal{S}$, *then* $\mathsf{RSAT}(\mathcal{S})$ *is NP-complete.*

Proof Sketch. Transformation of 3SAT to $\mathsf{RSAT}(\mathcal{S})$. $R_t = \{NTPP\}$ and $R_f = \{EQ\}$. Polarity constraints:

$$X_L\{DC, EC, TPP\}Y_{\neg L}, X_L\{DC, NTPP^{-1}\}X_{\neg L},$$
$$Y_L\{EC, TPP\}Y_{\neg L}, X_{\neg L}\{DC, EC, PO, TPP, NTPP\}Y_L$$

Clause constraints for each clause $c = \{i, j, k\}$:

$$Y_i\{NTPP^{-1}\}X_j, Y_j\{NTPP^{-1}\}X_k, Y_k\{NTPP^{-1}\}X_i.$$

All the above relations are contained in $\widehat{\mathcal{S}}$. Thus, $\mathsf{RSAT}(\widehat{\mathcal{S}})$ is NP-complete and because of Corollary 6.6, $\mathsf{RSAT}(\mathcal{S})$ is also NP-complete. ∎

Using these results, four other relations can be ruled out to be contained in any tractable subset of RCC-8, since the closure of the base relations plus one of the four relations contains $\{EQ, NTPP\}$ as well as $\{EC, TPP\}$.

Definition 7.13. *The two sets* $\mathcal{NP}_8$ *and* $\mathcal{P}_8$ *are defined as follows:*

- $\mathcal{NP}_8 = \mathcal{N}_1 \cup \mathcal{N}_2 \cup \{\{EC, NTPP, EQ\}, \{EC, NTPP^{-1}, EQ\},$
    $\{DC, EC, NTPP, EQ\}, \{DC, EC, NTPP^{-1}, EQ\}\}.$
- $\mathcal{P}_8 = \text{RCC-8} \setminus \mathcal{NP}_8.$

$\mathcal{NP}_8$ contains 76 relations, $\mathcal{P}_8$ contains 180 relations. Combining Lemma 6.11, Lemma 7.11, and Lemma 7.12 leads to the following corollary.

Corollary 7.14. $\mathsf{RSAT}(\mathcal{B} \cup \{R\})$ *is NP-hard, if* $R \in \mathcal{NP}_8$.

Lemmata 7.11 and 7.12 give us a sufficient but not necessary condition of whether a subset of $\mathcal{P}_8$ is NP-hard, namely, if $\mathcal{B} \subseteq \widehat{\mathcal{S}} \subseteq \mathcal{P}_8$ contains $\{EQ, NTPP\}$ and one of $\{EC, TPP\}$ or $\{EC, NTPP\}$, RSAT($\mathcal{S}$) is NP-hard, otherwise the complexity of RSAT($\mathcal{S}$) remains open.

We computed all subsets of $\mathcal{P}_8$ that are candidates for maximal tractable subsets with respect to the above two NP-hardness proofs, i.e., we enumerated all subsets $\mathcal{S} \subseteq \mathcal{P}_8$ with $\mathcal{S} = \widehat{\mathcal{S}}$ that contain all base relations and the universal relation, and checked whether they fulfill the following two properties:

1. $\mathcal{S}$ does not contain ($\{EQ, NTPP\}$ and $\{EC, TPP\}$) or ($\{EQ, NTPP\}$ and $\{EC, NTPP\}$), and
2. the closure of $\mathcal{S}$ plus any relation of $\mathcal{P}_8 \setminus \mathcal{S}$ contains ($\{EQ, NTPP\}$ and $\{EC, TPP\}$) or ($\{EQ, NTPP\}$ and $\{EC, NTPP\}$).

To our surprise, this resulted in only three different subsets of $\mathcal{P}_8$, of which $\widehat{\mathcal{H}}_8$ is of course one of them. The other two subsets are denoted by $\mathcal{C}_8$ and $\mathcal{Q}_8$.[2]

$$
\begin{aligned}
\widehat{\mathcal{H}}_8 \;=\; & \mathcal{P}_8 \setminus \{R \mid (\{EQ, NTPP\} \subseteq R \text{ and } \{TPP\} \not\subseteq R) \\
& \qquad \text{or } (\{EQ, NTPP^{-1}\} \subseteq R \text{ and } \{TPP^{-1}\} \not\subseteq R)\} \\
\mathcal{C}_8 \;=\; & \mathcal{P}_8 \setminus \{R \mid \{EC\} \subset R \text{ and } \{PO\} \not\subseteq R \text{ and} \\
& \qquad R \cap \{TPP, NTPP, TPP^{-1}, NTPP^{-1}, EQ\} \neq \emptyset\} \\
\mathcal{Q}_8 \;=\; & \mathcal{P}_8 \setminus \{R \mid \{EQ\} \subset R \text{ and } \{PO\} \not\subseteq R \text{ and} \\
& \qquad R \cap \{TPP, NTPP, TPP^{-1}, NTPP^{-1}\} \neq \emptyset\}
\end{aligned}
$$

$\mathcal{C}_8$ contains 158 different RCC-8 relations, $\mathcal{Q}_8$ contains 160 different relations. We have that $\widehat{\mathcal{H}}_8 \cup \mathcal{C}_8 = \mathcal{P}_8$, i.e., each of the 180 relations of $\mathcal{P}_8$ is contained in at least one of the two sets.

Lemma 7.15. *For every subset $\mathcal{S}$ of RCC-8: If RSAT($\mathcal{S}$) is tractable, then $\mathcal{S} \subseteq \widehat{\mathcal{H}}_8$, $\mathcal{S} \subseteq \mathcal{C}_8$, or $\mathcal{S} \subseteq \mathcal{Q}_8$.*

Proof. By computing the closure of every set containing the base relations, the universal relation, and two arbitrary RCC-8 relations, the resulting set is either contained in one of the sets $\widehat{\mathcal{H}}_8$, $\mathcal{C}_8$, or $\mathcal{Q}_8$, or is NP-hard according to Lemma 7.11, Lemma 7.12, or Corollary 7.14. ∎

So far we have not said anything about the tractability of $\mathcal{C}_8$ or $\mathcal{Q}_8$ or subsets thereof. It might be that both sets are NP-hard or that there is a large number of different maximal tractable subsets that are contained in the two sets. What we have obtained so far is, thus, only a restriction of the number of different subsets we have to check for tractability. However, we will see in the next section that both $\mathcal{C}_8$ and $\mathcal{Q}_8$ are in fact tractable.

---

[2] The names $\mathcal{C}_8$ and $\mathcal{Q}_8$ were chosen, since all $\mathcal{P}_8$-relations not contained in $\mathcal{C}_8$ contain E$\underline{C}$, and all $\mathcal{P}_8$-relations not contained in $\mathcal{Q}_8$ contain E$\underline{Q}$.

7.3 A Complete Analysis of Tractability

In Chapter 6 tractability of $\widehat{\mathcal{H}}_8$ was shown by reducing $\mathsf{RSAT}(\widehat{\mathcal{H}}_8)$ to HORN-SAT. This was possible because $\widehat{\mathcal{H}}_8$ contains exactly those relations that are transformed to propositional Horn formulas when using the propositional encoding of RCC-8. This propositional Horn encoding of $\widehat{\mathcal{H}}_8$ was also used for proving that path-consistency decides $\mathsf{RSAT}(\widehat{\mathcal{H}}_8)$, by relating positive unit resolution, which is a complete decision method for propositional Horn formulas, to path-consistency. The propositional encoding of $\mathcal{C}_8$ and $\mathcal{Q}_8$ is neither a Horn formula nor a Krom formula (two literals per clause), so it is not immediately possible to reduce the consistency problem of these subsets to a tractable propositional satisfiability problem. Therefore we have to find other ways of proving tractability.

Let us have a closer look at the relations of the two sets $\mathcal{C}_8$ and $\mathcal{Q}_8$ and how they differ from $\widehat{\mathcal{H}}_8$.

Proposition 7.16. *For all relations $R \in \mathcal{P}_8 \setminus \widehat{\mathcal{H}}_8$ we have that $\{\mathsf{EQ}\} \subset R$. For all refinements R' of R with $R' \cup \{\mathsf{EQ}\} = R$ and $\{\mathsf{EQ}\} \not\subseteq R'$ we have that $R' \in \widehat{\mathcal{H}}_8$.*

$\mathcal{C}_8$ and $\mathcal{Q}_8$ are both subsets of $\mathcal{P}_8$, so if we can prove the following conjecture, then both sets are tractable.

Conjecture 7.17. *Let Θ be a path-consistent set of constraints over $\mathcal{C}_8$ or over $\mathcal{Q}_8$. If Θ' is obtained from Θ by eliminating the identity relation $\{\mathsf{EQ}\}$ from every constraint $xRy \in \Theta$ with $R \in \mathcal{P}_8 \setminus \widehat{\mathcal{H}}_8$, then enforcing path-consistency to Θ' does not result in an inconsistency.*

If we can prove this conjecture, then, by Proposition 7.16, Θ' contains only constraints over $\widehat{\mathcal{H}}_8$. Since path-consistency decides $\mathsf{RSAT}(\widehat{\mathcal{H}}_8)$, we then have that if Θ is path-consistent, Θ' is consistent and, therefore, Θ is also consistent, i.e., path-consistency decides consistency for sets of constraints over $\mathcal{C}_8$ and over $\mathcal{Q}_8$. In Section 6.5 it was shown that the relation $\{\mathsf{EQ}\}$ can always be eliminated from every constraint of a path-consistent set of constraints over $\widehat{\mathcal{H}}_8$ without changing consistency of the set. This was, again, shown by applying positive unit resolution to the propositional Horn encoding of $\widehat{\mathcal{H}}_8$, a method which is not applicable in our case. Instead, we can now apply our new method developed in Section 7.1, namely, we can check whether $\mathcal{C}_8$ and $\mathcal{Q}_8$ can be reduced by refinement to $\widehat{\mathcal{H}}_8$. Conjecture 7.17 gives the refinement matrix we have to check. We define this particular refinement matrix for arbitrary sets $\mathcal{S}$ of disjunctions over a set $\mathcal{A}$ of JEPD relations:

Definition 7.18 (identity-refinement matrix).
$M^{\neq}$ is the identity-refinement matrix of a set $\mathcal{S} \subseteq 2^{\mathcal{A}}$ if for every $S \in \mathcal{S}$, $M^{\neq}[S][S'] = true$ iff $S' = S \setminus Id$, where $Id \in \mathcal{A}$ is the identity relation.

Using the identity-refinement matrix it turns out that both $\mathcal{C}_8$ and $\mathcal{Q}_8$ can be reduced by refinement to $\widehat{\mathcal{H}}_8$.

Proposition 7.19.

- CHECK-REFINEMENTS$(\mathcal{C}_8, M^{\neq})$ *returns* succeed.
- CHECK-REFINEMENTS$(\mathcal{Q}_8, M^{\neq})$ *returns* succeed.

Theorem 7.20. *Path-consistency decides* $\mathsf{RSAT}(\mathcal{C}_8)$ *as well as* $\mathsf{RSAT}(\mathcal{Q}_8)$.

Proof. It follows from Proposition 7.16 and Proposition 7.19 that Theorem 7.5 can be applied with $\mathcal{T} = \widehat{\mathcal{H}}_8$. Since path-consistency decides $\mathsf{RSAT}(\widehat{\mathcal{H}}_8)$, it follows from Corollary 7.6 that path-consistency also decides $\mathsf{RSAT}(\mathcal{C}_8)$ as well as $\mathsf{RSAT}(\mathcal{Q}_8)$. ∎

Together with Lemma 7.15 it follows that $\widehat{\mathcal{H}}_8$, $\mathcal{C}_8$, and $\mathcal{Q}_8$ are the only maximal tractable subsets of RCC-8 that contain all base relations.

Theorem 7.21. *For every subset $\mathcal{S}$ of* RCC-8 *that contains all base relations and the universal relation:* $\mathsf{RSAT}(\mathcal{S})$ *is tractable iff $\mathcal{S} \subseteq \widehat{\mathcal{H}}_8$, $\mathcal{S} \subseteq \mathcal{C}_8$, or $\mathcal{S} \subseteq \mathcal{Q}_8$.*

In Appendix A.3 we list all relations of RCC-8 and give their membership in the three maximal tractable subsets.

7.4 Finding a Consistent Scenario II: An Improved Algorithm for All Tractable Subsets

In Section 6.5 we gave an $O(n^3)$ time algorithm for computing a consistent scenario for a set of constraints over $\widehat{\mathcal{H}}_8$. This algorithm is based on first eliminating $\{\mathsf{EQ}\}$ from every constraint and then successively refining constraints to constraints over base relations in a particular order and enforcing path-consistency in between. As shown in the previous section it is also possible to eliminate $\{\mathsf{EQ}\}$ from all constraints over $\mathcal{C}_8$ and over $\mathcal{Q}_8$. Since all resulting constraints are over $\widehat{\mathcal{H}}_8$ relations (see Proposition 7.16), it is possible to apply our previous algorithm also for $\mathcal{C}_8$ and $\mathcal{Q}_8$. By applying the method developed in Section 7.1 we can, however, improve this algorithm for all three maximal tractable subsets of RCC-8.

When analyzing the resulting refinement matrices of Proposition 7.19, we found that for almost all relations one of the entries in the refinement matrix is a base relation which is related to the original relation according to a particular refinement scheme. If all relations had a base relation as an entry in the refinement matrix, then finding a consistent scenario would be trivial, namely, by just replacing each relation with the corresponding base relation. We, thus, generated refinement arrays $A_{\mathcal{C}}$ and $A_{\mathcal{Q}}$ for the maximal tractable subsets $\mathcal{C}_8$ and $\mathcal{Q}_8$, respectively, in which we insert the base relations of the corresponding refinement matrices. For those relations that do not have a base relation as an entry in the refinement matrix, we assign a base relation according to the identified refinement scheme. This refinement scheme is similar to the one identified in Section 6.5 for $\widehat{\mathcal{H}}_8$. According to Theorem 6.63, we also generated a refinement array $A_{\mathcal{H}}$ for $\widehat{\mathcal{H}}_8$. The refinement schemes for the three maximal tractable subsets are given in the following proposition.

Proposition 7.22. *Let $A_{\mathcal{H}}, A_{\mathcal{C}},$ and $A_{\mathcal{Q}}$ be the refinement arrays corresponding to the sets $\widehat{\mathcal{H}}_8, \mathcal{C}_8,$ and $\mathcal{Q}_8$, respectively, and let $\mathcal{S}$ be one of $\widehat{\mathcal{H}}_8, \mathcal{C}_8,$ or $\mathcal{Q}_8$. $A_{\mathcal{S}}$ is obtained according to the following refinement scheme:*
For every relation $R \in \mathcal{S}$,

    1. if $R \in \mathcal{B}$, then $A_{\mathcal{S}}(R) = R$;
    2. else if $\{DC\} \subseteq R$, then $A_{\mathcal{S}}(R) = \{DC\}$;
    3. else if $\{EC\} \subseteq R$ and $\mathcal{S} = \mathcal{Q}_8$ or $\mathcal{S} = \widehat{\mathcal{H}}_8$, then $A_{\mathcal{S}}(R) = \{EC\}$;
    4. else if $\{PO\} \subseteq R$, then $A_{\mathcal{S}}(R) = \{PO\}$;
    5. else if $\{NTPP\} \subseteq R$ and $\mathcal{S} = \mathcal{C}_8$, then $A_{\mathcal{S}}(R) = \{NTPP\}$;
    6. else if $\{TPP\} \subseteq R$, then $A_{\mathcal{S}}(R) = \{TPP\}$;
    7. else $A_{\mathcal{S}}(R) = A_{\mathcal{S}}(R^{\smile})$.

It can be easily verified that the above refinement schemes assigns a base relation to each relation of the maximal tractable subsets. Thus, if these refinement arrays are valid, all three maximal tractable subsets can be reduced by refinement to $\mathcal{B}$, the set of all RCC-8 base relations.

Proposition 7.23. *If $\mathcal{S}$ is one of $\widehat{\mathcal{H}}_8, \mathcal{C}_8,$ or $\mathcal{Q}_8$, then* CHECK-REFINE-MENTS-$2(\mathcal{S}, A_{\mathcal{S}})$ *returns* succeed.

All three refinement arrays are valid, so this gives us the possibility of computing a consistent scenario of a path-consistent set of constraints over a tractable subset by a simple table lookup scheme, which is the fastest possible way.

Theorem 7.24. *A consistent scenario Θ_s of a path-consistent set Θ of constraints over $\widehat{\mathcal{H}}_8, \mathcal{C}_8,$ or over $\mathcal{Q}_8$ can be computed in $O(n^2)$ time, by replacing every constraint $xRy \in \Theta$ with $xR'y \in \Theta_s$, where $R' = A_{\mathcal{H}}(R), A_{\mathcal{C}}(R),$ or $A_{\mathcal{Q}}(R)$, respectively.*

7.5 Applying the New Method to Allen's Interval Algebra

As a way of demonstrating generality of our method developed in Section 7.1, we apply it to Allen's interval algebra [4]. Since the only maximal tractable subclass of the interval algebra containing all base relations has already been identified [128], we cannot provide any new results on that topic. We can, however, validate the usefulness of our method by showing that tractability of the ORD-Horn subclass $\mathcal{H}$ and sufficiency of path-consistency for deciding consistency for this subclass can also be proven using our method. For this we apply the same strategy as for $\mathcal{C}_8$ and $\mathcal{Q}_8$, namely, we use the identity-refinement matrix $M^{\neq}$ (cf. [115]) and reduce $\mathcal{H}$ by refinement to the pointisable subclass $\mathcal{P}$. Since the number of relations of Allen's interval algebra is much larger than the number of relations of RCC-8, we use this time our second algorithm CHECK-REFINEMENTS-2 and an *identity-refinement array* instead of an identity-refinement matrix which is defined as follows:

Definition 7.25 (identity-refinement array).
$A^{\neq}$ *is the* identity-refinement array *of a set* $\mathcal{S} \subseteq 2^{\mathcal{A}}$ *if for every* $S \in \mathcal{S}$, $A^{\neq}[S] = S \setminus Id$, *where* $Id \in \mathcal{A}$ *is the identity relation.*

Running our method using the identity-refinement array, we obtain the following result.

Proposition 7.26. GET-REFINEMENTS-2$(\mathcal{H}, A^{\neq})$ *returns* $A^{\mathcal{P}}$ *which has the following property:* $A^{\mathcal{P}}[R] \in \mathcal{P}$ *for every* $R \in \mathcal{H}$.

This was computed in less than 25 minutes on a Sun Sparc Ultra1. We can now apply Theorem 7.9:

Lemma 7.27. *The ORD-Horn subclass* $\mathcal{H}$ *can be reduced by refinement to the pointisable subclass* $\mathcal{P}$.

Since path-consistency is sufficient for deciding inconsistency of reasoning over pointisable relations [156], it follows from Lemma 7.2 that the same is true for reasoning over ORD-Horn relations. This gives us a very simple and automatically generated proof of the results obtained by Nebel and Bürckert [128].

Nevertheless, when analyzing the resulting refinement array $A^{\mathcal{P}}$, we also obtained a new result about Allen's interval algebra. Not all of the 188 pointisable relations are used in $A^{\mathcal{P}}$, but only 30 of them which are obtained according to the following refinement scheme.

Proposition 7.28. *For every relation* $R \in \mathcal{H}$, *the corresponding pointisable relation* $point(R)$ *is obtained according to the following refinement scheme* $(R^{\neq} = R \setminus \{=\}$ *and* $R' = R \cap \{<, >, o, oi, d, di\})$:

1. *If* $R = \{=\}$, *then* $point(R) = R$;
2. *else if* $R' \neq \{\}$, *then* $point(R) = R'$;
3. *else* $point(R) = R^{\neq}$.

Using this refinement scheme, every path-consistent set Θ of constraints over ORD-Horn can be refined in $O(n^2)$ time to a path-consistent set Θ' by replacing every constraint $xRy \in \Theta$ with $xR'y \in \Theta'$, where $R' = point(R)$. This can be useful for some tasks such as finding a consistent scenario (cf. [67, 155]).

7.6 Discussion & Further Work

We developed a general method for proving sufficiency of path-consistency for deciding consistency of disjunctions over a set of JEPD relations. We applied this method to the Region Connection Calculus RCC-8 and identified two large new maximal tractable subsets. Together with $\widehat{\mathcal{H}}_8$, the already known maximal tractable subset, these are the only such sets for RCC-8 that contain all base relations and can, hence, be used to increase efficiency of backtracking algorithms for reasoning over NP-hard subsets. We also applied our method to Allen's interval algebra which resulted in a simple proof of tractability of reasoning in the ORD-Horn subclass.

Our new method is very simple, does not require detailed knowledge about the considered relations, and seems to be very powerful. The only difficulty of this method is finding an adequate refinement matrix, resp. refinement array—the number of different refinement matrices/arrays is exponential in the number of relations. However, the simple heuristic of eliminating all identity relations was successful for all maximal tractable fragments of RCC-8 and Allen's interval algebra which contain all base relations. This leads to an interesting question, namely, whether it is a general property of sets of relations containing all base relations for which path-consistency decides consistency that the identity relation can be eliminated from all constraints of a path-consistent set without making the set inconsistent.

Furthermore, the identity-refinement matrix/array was also very effective in identifying intractable subsets. For instance, we first identified candidates for maximal tractable subsets of RCC-8 without using the additional NP-hardness proofs given in Lemmata 7.11 and 7.12. For these candidates, CHECK-REFINEMENTS immediately returned `fail`. Analyzing which relations lead to the inconsistencies helped us in finding the NP-hard subsets of Lemmata 7.11 and 7.12.

Even if the identity-refinement matrix/array does not work or is not sufficient for reducing a set by refinement to another set, it is possible to add just a few refinements and run GET-REFINEMENTS. The algorithm either tells that the refinements are not valid or makes additional refinements. In this "greedy" manner, an adequate refinement matrix/array can be identified without having to check many different possibilities.

8. Empirical Evaluation of Reasoning with RCC-8

In the previous chapters we studied the computational properties of RCC-8. We found that reasoning over RCC-8 relations is NP-complete and identified three maximal tractable subsets of RCC-8 that contain all base relations. These sets can be used to speed up backtracking search for the general NP-complete reasoning problem by reducing the search space considerably (see Table 6.1). Nevertheless, the search space is still exponential in the size of the problem instance and it cannot be expected that it is possible to solve large instances in reasonable time.

In this chapter we tackle several questions that emerge from the results of the previous chapters. Up to which size is it possible to solve instances in reasonable time? Which heuristic is the best? Is it really so much more efficient to use the maximal tractable subsets for solving instances of the NP-complete consistency problem as Table 6.1 indicates or is this effect out-balanced by the forward-checking power of the interleaved path-consistency computations? This was the case for similar temporal problems (pointisable vs. ORD-Horn relations) [127]. Is it possible to combine the different heuristics in such a way that more instances can be solved in reasonable time than by each heuristic alone?

We treat these questions by randomly generating instances and solving them using different heuristics. In doing so, we are particularly interested in the hardest randomly generated instances which leads to the question of phase-transitions [20]: Is there a parameter for randomly generating instances of the consistency problem of RCC-8 that results in a phase-transition behavior? If so, is it the case that the hardest instances are mainly located in the phase-transition region while the instances not contained in the phase-transition region are easily solvable? In order to generate instances which are harder with a higher probability, we generate two different kinds of instances. On the one hand instances which contain constraints over all RCC-8 relations, on the other hand instances which contain only constraints over relations which are not contained in any of the maximal tractable subsets. We expect these instances to be harder on average than the former instances.

J. Renz: Qualitative Spatial Reasoning with Topological Information, LNAI 2293, pp. 131–154, 2002.
© Springer-Verlag Berlin Heidelberg 2002

8.1 Test Instances, Heuristics, and Measurement

There is no previous work on empirical evaluation of algorithms for reasoning with RCC-8 and no benchmark problems are known. Therefore we randomly generated our test instances with a given number of regions n, an average label-size l, and an average degree d of the constraint graph. Further, we used two different sets of relations for generating test instances, the set of all RCC-8 relations and the set of hard RCC-8 relations $\mathcal{NP}_8$, i.e., those 76 relations which are not contained in any of the maximal tractable subsets $\widehat{\mathcal{H}}_8$, $\mathcal{C}_8$, or $\mathcal{Q}_8$. Based on these sets of relations, we used two models to generate instances, denoted by $A(n, d, l)$ and $H(n, d, l)$. The former model uses all relations to generate instances, the latter only the relations in $\mathcal{NP}_8$. The instances are generated as follows:

1. A constraint graph with n nodes and an average degree of d for each node is generated. This is accomplished by selecting $nd/2$ out of the $n(n-1)/2$ possible edges using a uniform distribution.
2. If there is no edge between the ith and jth node, we set $R_{ij} = R_{ji}$ to be the universal relation.
3. Otherwise a non-universal relation is selected according to the parameter l such that the average size of relations for selected edges is l. This is accomplished by selecting one of the base relations with uniform distribution and out of the remaining 7 relations each one with probability $(l-1)/7$.[1] If this results in an allowed relation (i.e., a relation of $\mathcal{NP}_8$ for $H(n, d, l)$, any RCC-8 relation for $A(n, d, l)$), we assign this relation to the edge. Otherwise we repeat the process.

The reason for also generating instances using only relations of $\mathcal{NP}_8$ is that we assume that these instances are difficult to solve since every relation has to be split during the backtracking search, even if we use a maximal tractable subclass as the split-set. We generated only instances of average label size $l = 4.0$, since in this case the relations are equally distributed.

This way of generating random instances is very similar to the way random CSP instances over finite domains are usually generated (see, e.g., [64].) Achlioptas et al. [2] found that the standard models for generating random CSP instances over finite domains lead to trivially flawed instances for $n \rightarrow \infty$, i.e., instances become locally inconsistent without having to propagate constraints. Since we are using CSP instances over infinite domains, Achlioptas et al.'s result does not necessarily hold for our random instances. We, therefore, analyze in the following whether our instances are also trivially flawed for $n \rightarrow \infty$. In order to obtain a CSP over a finite domain, we first have to transform our constraint graph into its dual

---

[1] This method could result in the assignment of a universal constraint to a selected link, thereby changing the degree of the node. However, since the probability of getting the universal relation is very low, we ignore this in the following.

graph where each of the $n(n-1)/2$ edges R_{ij} of our constraint graph corresponds to a node in the dual graph. Moreover, each of the n variables of the constraint graph corresponds to $n-1$ edges in the dual graph, i.e., the dual graph contains $n(n-1)$ edges and $n(n-1)/2$ nodes. In the dual graph, each node corresponds to a variable over the eight-valued domain $\mathcal{D} = \{\mathsf{DC}, \mathsf{EC}, \mathsf{PO}, \mathsf{TPP}, \mathsf{TPP}^{-1}, \mathsf{NTPP}, \mathsf{NTPP}^{-1}, \mathsf{EQ}\}$. Ternary constraints over these variables are imposed by the composition table, i.e., the composition rules $R_{ij} \subseteq R_{ik} \circ R_{kj}$ must hold for all connected triples of nodes R_{ij}, R_{ik}, R_{kj} of the dual graph ($R_{ij} = R_{ji}$ for all i, j). There are $\binom{n}{3}$ $= n(n-1)(n-2)/6$ connected triples in the dual graph. The overall number of triples in the dual graph is $\binom{n(n-1)/2}{3}$. $nd/2$ unary constraints on the domain of the variables R_{ij} are given, i.e., unary constraints are given for $\binom{nd/2}{3}$ triples in the dual graph. Therefore, the expected number E_{CT}^n of connected triples for which unary constraints are given can be computed as

$$E_{CT}^n = \binom{n}{3} \cdot \binom{nd/2}{3} / \binom{n(n-1)/2}{3}.$$

For $n \to \infty$, the expected number of triples E_{CT}^∞ tends to $d^3/6$. For the instances generated according to the model $A(n, d, l)$, the probability that the unary constraints which are assigned to a triple lead to a local inconsistency is about $0,0036\%$ (only 58,989 out of the $255^3 = 16,581,375$ possible assignments are inconsistent.) Since one locally inconsistent triple makes the whole instance inconsistent, we are interested in the average degree d for which the expected number E_{IT}^n of locally inconsistent triples is equal to one. For the model $A(n, d, l)$ this occurs for a value of $d = 11.90$, and $E_{IT}^\infty = 0.5$ for $d = 9.44$. For $n = 100$, the expected number of locally inconsistent triples is one for $d = 13.98$, and $E_{IT}^{100} = 0.5$ for $d = 11.10$. For the model $H(n, d, l)$, none of the possible assignments of the triples leads to a local inconsistency, i.e., all triples of the randomly generated instances of the $H(n, d, l)$ model are locally consistent.[2] This analysis shows that contrary to what Achlioptas et al. found for randomly generated CSP instances over finite domains, the model $H(n, d, l)$, and the model $A(n, d, l)$ for d small do not suffer from trivial local inconsistencies.

In order to solve the randomly generated instances of RSAT, we used the backtracking algorithm of Figure 2.3 which is based on the algorithm proposed by Ladkin and Reinefeld [111] for solving qualitative temporal reasoning problems [127]. This algorithm uses path-consistency for forward checking and backtracks on decompositions of the specified relations into sub-relations that permit the path-consistency algorithm to decide the consistency problem. The search space on which backtracking is performed depends on the

---

[2] This is similar to the result for CSPs over finite domains that by restricting the constraint type, e.g., if only "not-equal" constraints as in graph-coloring are used, it is possible to ensure that problems cannot be trivially flawed.

split-set, i.e., the set of sub-relations that is allowed in the decompositions. Choosing the right split-set influences the search noticeably as it influences the average branching factor of the search space. We choose five different split sets, the three maximal tractable subsets $\widehat{\mathcal{H}}_8$, $\mathcal{Q}_8$, and $\mathcal{C}_8$, the set of base relations $\mathcal{B}$ and the closure of this set $\widehat{\mathcal{B}}$ which consists of 38 relations. These sets have the following branching factors $\mathcal{B}$: 4.0, $\widehat{\mathcal{B}}$: 2.50 ,$\widehat{\mathcal{H}}_8$: 1.438, $\mathcal{C}_8$: 1.523, $\mathcal{Q}_8$: 1.516. This is, of course, a worst case measure because the interleaved path-consistency computations reduce the branching factor considerably [112].

Apart from the choice of the split-set there are other heuristics which influence the efficiency of the search. In general it is the best search strategy to proceed with the constraint with the most constraining relation (line 3 of Figure 2.3) and the least constraining choice of a sub-relation (line 5 of Figure 2.3.) We investigated two different strategies for choosing the next constraint to be processed [127].

Static/dynamic: Constraints are processed according to a heuristic evaluation of their constrainedness which is determined *statically* before the backtracking starts or *dynamically* during the search.

Local/global: The evaluation of the constrainedness is based on a *local* heuristic weight criterion or on a *global* heuristic criterion [157].

This gives us four possibilities we can combine with the five different split-sets, i.e., a total number of 20 different heuristics. The evaluation of constrainedness as well as how relations are decomposed into relations of different split-sets depends on the *restrictiveness* of relations which is a heuristic criterion [157]. Restrictiveness of a relation is a measure of how a relation restricts its neighborhood. For instance, the universal relation given in a constraint network does not restrict its neighboring relations at all, the result of the composition of any relation with the universal relation is the universal relation. The identity relation, in contrast, restricts its neighborhood a lot. In every triple of variables where one relation is the identity relation, the other two relations must be equal. Therefore, the universal relation is usually the least restricting relation, while the identity relation is usually the most restricting relation. Restrictiveness of relations is represented as a weight in the range of 1 to 16 assigned to every relation, where 1 is the value of the most and 16 the value of the least restricting relation. We discuss in the following section in detail how the restrictiveness and the weight of a relation is determined.

Given the weights assigned to every relation, we compute decompositions and estimate constrainedness as follows. For each split-set $\mathcal{S}$ and for each RCC-8 relation R we compute the smallest decomposition of R into sub-relations of $\mathcal{S}$, i.e., the decomposition which requires the least number of sub-relations of $\mathcal{S}$. If there is more than one possibility, we chose the decomposition with the least restricting sub-relations. In line 5 of the backtracking algorithm (see Figure 2.3), the least restricting sub-relation of each decomposition is processed first. For the local strategy, the constrainedness of a constraint is determined by the size of its decomposition (which can be different for every split-set) and by its weight. We choose the constraint with the smallest decomposition larger than one and, if there is more than one such constraint, the one with the smallest weight. The reason for choosing the relation with the smallest decomposition is that it is expected that forward-checking refines relations with a larger decomposition into relations with a smaller decomposition. This reduces the backtracking effort. For the global strategy, the constrainedness of a constraint xRy is determined by adding the weights of all neighboring relations S, T with xSz and zTy to the weight of R. The idea behind this strategy is that when refining the relation R with the most restricted neighborhood, an inconsistency is detected faster than when refining a relation with a less restricted neighborhood.

In order to evaluate the quality of the different heuristics, we measured the run-time used for solving instances as well as the number of visited nodes in the search space. Comparing different approaches by their run-time is often not very reliable as it depends on several factors such as the implementation of the algorithms, the used hardware, or the current load of the used machine which makes results sometimes not reproducible. For this reason, we ran all our run-time experiments on the same machine, a Sun Ultra 1 with 128MB of main memory. Nevertheless, we suggest to use the run-time results mainly for qualitatively comparing different heuristics and for getting a rough idea of the order of magnitude by which instances can be solved. In contrast to this, the number of visited nodes for solving an instance with a particular heuristic is always the same on every machine. This allows one to compare the path through the search space taken by the single heuristics and to judge which heuristic makes the better choices on average. However, this does not take into account the time that is needed to make a choice at a single node. Computing the local constrainedness of a constraint is certainly faster than computing its global constrainedness. Similarly, computing constrainedness statically should be faster than computing it dynamically. Furthermore, larger instances require more time at the nodes than smaller instances, be it for computing path-consistency or for computing the constrainedness. Taking running-time and the number of visited nodes together gives good indications of the quality of the heuristics.

A further choice we make in evaluating our measurements is that of how to aggregate the measurements of the single instances to a total picture. Some possibilities are to use either the average or different percentiles such as the median, i.e., the 50% percentile. The $d\%$ percentile for a value $0 < d < 100$ is obtained by sorting the measurements in increasing order and picking the measurement of the $d\%$ element, i.e., $d\%$ of the values are less than that value. Suppose that most instances have a low value (e.g. running time) and only a few instances have a very large value. Then the average might be larger than the values of almost all instances, while in this case the median is a better indication of the distribution of the values. In this case the 99% percentile, for instance, gives a good indication of the value of the hardest among the "normal" instances. We have chosen to use the average value when the measurements are well distributed and to use both 50% and 99% percentile when there are only a few exceptional values in the distribution of the measurements.

8.2 Empirical Evaluation of the Path-Consistency Algorithm

Since the efficiency of the backtracking algorithm depends on the efficiency of the underlying path-consistency algorithm, we will first compare different implementations of the path-consistency algorithm. In previous empirical investigations [157] of reasoning with Allen's interval relations [4], different methods for computing the composition of two relations were evaluated. This was mainly because the full composition table for the interval relations contains $2^{13} \times 2^{13} = 67108864$ entries which was too large at that time to be stored in the main memory. In our setting, we simply use a composition table that specifies the compositions of all RCC-8 relations, which is a 256×256 table consuming approximately 128kB of main memory. This means that the composition of two arbitrary relations is done by a simple table lookup.

Van Beek and Manchak [157] also studied the effect of weighting the relations in the queue according to their restrictiveness and process the most restricting relation first. Restrictiveness was measured for each base relation by successively composing the base relation with every possible label, summing up the cardinalities, i.e., the number of base relations contained in the result of the composition, and suitably scaling the result. The reason for doing so is that the most restricting relation restricts the other relations on average most and therefore decreases the probability that they have to be processed again. Restrictiveness of a complex relation was approximated by summing up the restrictiveness of the involved base relations. Van Beek and Manchak [157] found that their method of weighting the triples in the queue is much more efficient than picking randomly an arbitrary triple. Because of the relatively small number of RCC-8 relations, we computed the exact restrictiveness by composing each relation with every other relation and summing up the cardinalities of the resulting compositions. We scaled the result into weights from 1 (the most restricting relation) to 16 (the least restricting relations.)

This gives us three different implementations of the path-consistency algorithm given in Figure 2.2. One in which the entries in the queue are not weighted, one with approximated restrictiveness as done by van Beek and Manchak, and one with exact restrictiveness.[3] In order to compare these implementations, we randomly generated instances with 50 to 1,000 regions. For

---

[3] For the weighted versions we select a path (i, k, j) from the queue Q in line 3 of the algorithm of Figure 2.2 according to the weights of the different paths in Q which are computed as specified above.

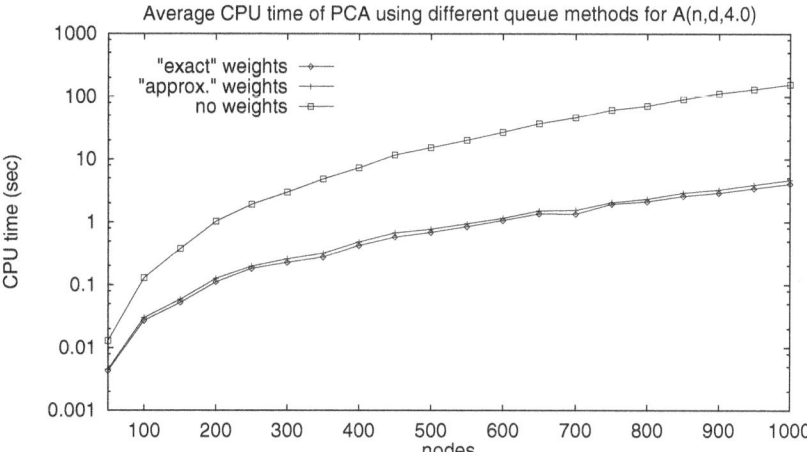

Fig. 8.1. Comparing the performance of the path-consistency algorithm using different methods for weighting the queue (70 instances/data point, $d = 8.0 - 11.0$)

each value of the average degree ranging from 8.0 stepping with 0.5 to 11.0 we generated 10 different instances. Figure 8.1 displays the average CPU time of the different methods for applying the path-consistency algorithm to the same generated instances. It can be seen that the effect of using a weighted queue is much greater for our problem than for the temporal problem (about 10× faster than using an ordinary queue without weights compared to only about 2× faster [157]). Determining the weights of every relation using their exact restrictiveness does not have much advantage over approximating their restrictiveness using the approach by van Beek and Manchak [157], however. For our further experiments we always used the "exact weights" method because determining the restrictiveness amounts to just one table lookup.

As mentioned in the previous section, one way of measuring the quality of the heuristics is to count the number of visited nodes in the backtrack search. In our backtracking algorithm, path-consistency is enforced in every visited node. Note that it is not adequate to multiply the average running-time for enforcing path-consistency of an instance of a particular size with the number of visited nodes in order to obtain an approximation of the required running time for that instance. The average running-time for enforcing path-consistency as given in Figure 8.1 holds only when all possible paths are entered into the queue at the beginning of the computation (see line 1 of Figure 2.2). These are the paths which have to be checked by the algorithm. The path-consistency computation during the backtracking search is different, however. There, only the paths involving the currently changed constraint are entered in the queue, since only these paths might result in changes of the constraint graph. This is much faster than the full compu-

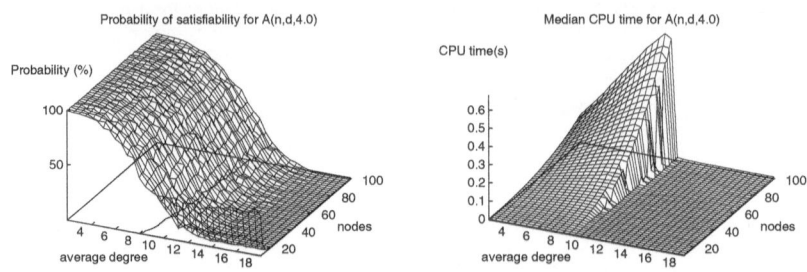

Fig. 8.2. Probability of satisfiability and median CPU time for $A(n, d, 4.0)$ using the $\widehat{\mathcal{H}}_8$/static/global heuristic (500 instances per data point)

tation of path-consistency which is only done once at the beginning of the backtrack search.

8.3 The Phase-Transition of RCC-8

In order to study the quality of different heuristics and algorithms with randomly generated instances of an NP-complete problem, it is very important to be aware of the phase-transition behavior [20] of the problem (see Section 2.3.2). This is because instances which are not contained in the phase-transition region are often very easily solvable by most algorithms and heuristics and are, thus, not very useful for comparing their quality. Conversely, hard instances which are better suited for comparing the quality of algorithms and heuristics are usually found in the phase-transition region.

In this section we identify the phase-transition region of randomly generated instances of the RSAT problem, both for instances using all RCC-8 relations and for instances using only relations of $\mathcal{NP}_8$. Similarly to the empirical analysis of qualitative temporal reasoning problems [127], it turns out that the phase-transition depends most strongly on the average degree d of the nodes in the constraint graph. If all relations are allowed, the phase-transition is around $d = 8$ to $d = 10$ depending on the instance size (see Figure 8.2). Because of the result of our theoretical analysis of the occurrence of trivial flaws (see Section 8.1), it can be expected that for larger instance sizes the phase-transition behavior will be overlaid and mainly determined by the expected number of locally inconsistent triples which also depends on the average degree d. Thus, although it seems that the phase-transition shifts towards larger values of d as the instance size increases, the phase-transition asymptotically approaches $d = 9.44$, the theoretical value for $n \to \infty$ (see Section 8.1). Instances which are not path-consistent can be solved very fast by just one application of the path-consistency algorithm without further need for backtracking. When looking at the median CPU

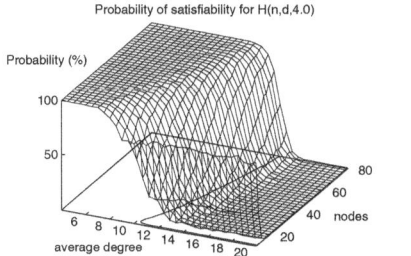
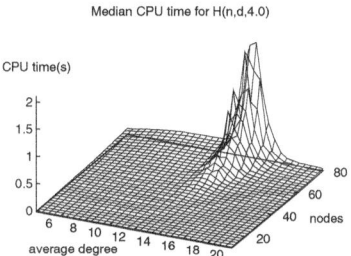

Fig. 8.3. Probability of satisfiability and median CPU time for $H(n, d, 4.0)$ using the $\widehat{\mathcal{H}}_8$/static/global heuristic (500 instances per data point)

times given in Figure 8.2, one notices that there is a sharp decline of the median CPU times at the phase transition. This indicates that for values of the average degree which are higher than where the phase-transition occurs, at least 50% of the instances are not path-consistent.

When using only "hard" relations, i.e., relations in $\mathcal{NP}_8$, the phase-transition appears at higher values for d, namely, between $d = 10$ and $d = 15$ (see Figure 8.3). As the median runtime shows, these instances are much harder in the phase-transition than in the former case. As in the previous case, but even stronger, it seems that the phase-transition shifts towards larger values of d as the instance size increases, and also that the phase-transition region narrows.

In order to evaluate the quality of the path-consistency method as an approximation to consistency, we counted the number of instances that are inconsistent but path-consistent (see Figure 8.4), i.e., those instances where the approximation of the path-consistency algorithm to consistency is wrong. First of all, one notes that all such instances are close to the phase transition region. In the general case, i.e., when constraints over all RCC-8 relations are employed, only a very low percentage of instances are path-consistent but inconsistent. Therefore, the figure looks very erratic. More data points would be required in order to obtain a smooth curve. However, a few important observations can be made from this figure, namely, that path-consistency gives an excellent approximation to consistency even for instances of a large size. Except for a very few instances in the phase-transition region, almost all instances which are path-consistent are also consistent. This picture changes when looking at the $H(n, d, 4.0)$ case. Here almost all instances in the phase-transition region and many instances in the mostly insoluble region are path-consistent, though only a few of them are consistent.

For the following evaluation of the different heuristics we will randomly generate instances with an average degree between $d = 2$ and $d = 18$ in the $A(n, d, 4.0)$ case and between $d = 4$ and $d = 20$ in the $H(n, d, 4.0)$ case. This covers a large area around the phase-transition. We expect the instances in

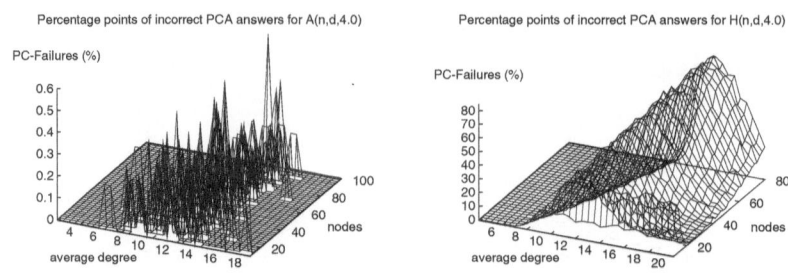

Fig. 8.4. Percentage points of incorrect answers of the path-consistency algorithm for $A(n, d, 4.0)$ and $H(n, d, 4.0)$

the phase-transition region of $H(n, d, 4.0)$ to be particularly hard which makes them very interesting for comparing the quality of the different heuristics.

8.4 Empirical Evaluation of the Heuristics

In this section we compare the different heuristics by running them on the same randomly generated instances. For the instances of $A(n, d, 4.0)$ we ran all 20 different heuristics (*static/dynamic* and *local/global* combined with the five split-sets $\mathcal{B}, \widehat{\mathcal{B}}, \widehat{\mathcal{H}}_8, \mathcal{C}_8, \mathcal{Q}_8$) on the same randomly generated instances of size $n = 10$ up to $n = 100$. For the instances of $H(n, d, 4.0)$ we restricted ourselves to instances with up to $n = 80$ regions because larger ones appeared to be too difficult.

In first experiments we found that most of the instances were solved very fast with less than 1,000 visited nodes in the search space when using one of the maximal tractable subsets for splitting. However, some instances turned out to be extremely hard, they could not be solved within our limit of 2 million visited nodes, which is about 1.5 hours of CPU time. Therefore, we ran all our programs up to a maximal number of 10,000 visited nodes and stored all instances for which at least one of the different heuristics used more than 10,000 visited nodes for further experiments (see next section). We call those instances the *hard instances*. The distribution of the hard instances is shown in Figure 8.5. It turned out that for the heuristics using $\mathcal{B}$ as a split set and for the heuristics using *dynamic* and *global* evaluation of the constrainedness many more instances were hard than for the other combinations of heuristics. We, therefore, did not include in Figure 8.5 the hard instances of the $\mathcal{B}/dynamic/global$ heuristic for $A(n, d, 4.0)$ and the hard instances for the heuristics using $\mathcal{B}$ as a split-set and the $\widehat{\mathcal{B}}/dynamic/global$ heuristic for $H(n, d, 4.0)$.

As Figure 8.5 shows, almost all of the hard instances are in the phase-transition region. For $A(n, d, 4.0)$ only a few of the 500 instances per data

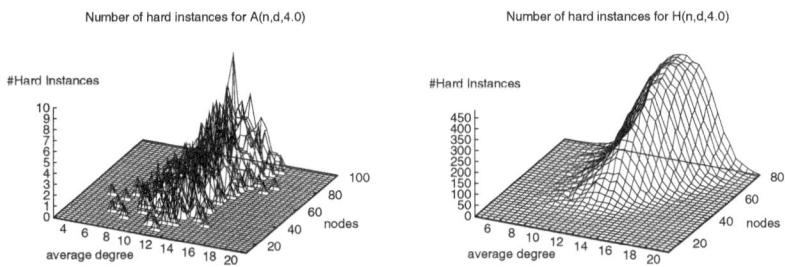

Fig. 8.5. Number of instances using more than 10,000 visited nodes for some heuristic for $A(n, d, 4.0)$ and $H(n, d, 4.0)$

point are hard while for $H(n, d, 4.0)$ almost all instances in the phase-transition are hard. Altogether there are 788 hard instances for $A(n, d, 4.0)$ (out of a total number of 759,000 generated instances) and 75,081 hard instances for $H(n, d, 4.0)$ (out of a total number of 594,000 generated instances). Table 8.1 shows the number of hard instances for each heuristic except for those which were excluded as mentioned above. The heuristics using $\widehat{\mathcal{H}}_8$ as a split-set solve more instances than the heuristics using other split-sets. Using $\mathcal{C}_8$ or $\mathcal{Q}_8$ as a split-set does not seem to be an improvement over using $\widehat{\mathcal{B}}$. Among the different ways of computing constrainedness, *static* and *global* appears to be the most effective combination when using one of the maximal tractable subsets as a split-set. For some split-sets, *dynamic* and *local* also seems to be an effective combination while combining *dynamic* and *global* is in all cases the worst choice with respect to the number of solved instances.

In Figure 8.6 we compare the 50% and 99% percentiles of the different heuristics on $A(n, d, 4.0)$. We do not give the average run times since we ran all heuristics only up to at most 10,000 visited nodes which reduces the real average run time values. Each data point is the average of the values for $d = 8$ to $d = 10$. We took the average of the different degrees in order to cover the whole phase-transition region which is about $d = 8$ for instances of size $n = 10$ and $d = 10$ for instances of size $n = 100$. For all different combinations of computing constrainedness, the ordering of the run times is the same for the different split-sets: $\mathcal{B} \gg \widehat{\mathcal{B}} > \mathcal{C}_8, \widehat{\mathcal{H}}_8, \mathcal{Q}_8$. The run times of using *static/local*, *static/global*, or *dynamic/local* for computing constrainedness are almost the same when combined with the same split-set while they are longer for all split-sets when using *dynamic/global* (about 3 times longer when using $\widehat{\mathcal{B}}$ as a split-set and about 1.5 times longer when using the other split-sets). The 99% percentile run times are only about 1.5 times longer than the 50% percentile run times. Thus, even the harder among the "normal" instances can be solved easily, i.e., apart from a few hard instances, most instances can be solved efficiently within the size range we analyzed. The erratic behavior of the median curves results from an aggregation of the effect which we already

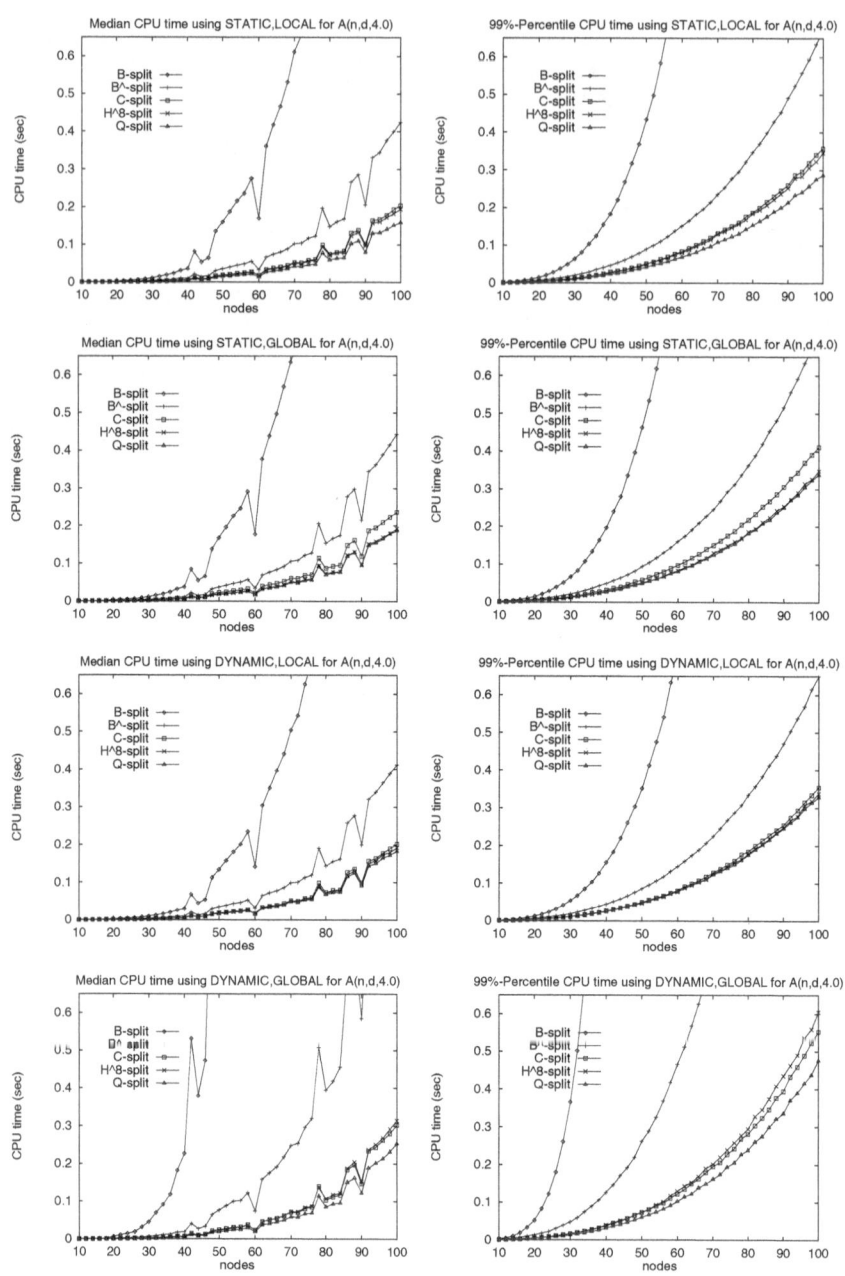

Fig. 8.6. Percentile 50% and 99% CPU time of the different heuristics for solving $A(n, d, 4.0)$ ($d = 8.0$ to $d = 10.0$, 2,500 instances per data point)

Table 8.1. Number of hard instances for each heuristic

Heuristics	$A(n,d,4.0)$	$H(n,d,4.0)$	$H(80,14.0,4.0)$
$\widehat{\mathcal{H}}_8$/sta/loc	64	21,129	331
$\widehat{\mathcal{H}}_8$/sta/glo	42	10,826	227
$\widehat{\mathcal{H}}_8$/dyn/loc	52	9,967	217
$\widehat{\mathcal{H}}_8$/dyn/glo	100	24,038	345
$\mathcal{C}_8$/sta/loc	81	28,830	373
$\mathcal{C}_8$/sta/glo	58	15,457	277
$\mathcal{C}_8$/dyn/loc	78	32,926	412
$\mathcal{C}_8$/dyn/glo	108	41,565	428
$\mathcal{Q}_8$/sta/loc	81	24,189	346
$\mathcal{Q}_8$/sta/glo	54	13,189	239
$\mathcal{Q}_8$/dyn/loc	74	13,727	255
$\mathcal{Q}_8$/dyn/glo	104	29,448	368
$\widehat{\mathcal{B}}$/sta/loc	68	23,711	344
$\widehat{\mathcal{B}}$/sta/glo	89	13,831	249
$\widehat{\mathcal{B}}$/dyn/loc	70	29,790	379
$\widehat{\mathcal{B}}$/dyn/glo	162	–	–
$\mathcal{B}$/sta/loc	163	–	–
$\mathcal{B}$/sta/glo	222	–	–
$\mathcal{B}$/dyn/loc	209	–	–
$\mathcal{B}$/dyn/glo	(303)	–	–
total	788	75,081	486

observed in Figure 8.2, namely, that some of the median elements in the phase-transition are inconsistent and easily solvable.

For the runtime studies for $H(n,d,4.0)$ we noticed that there are many hard instances for $n > 40$ (see Figure 8.5), for $n = 80$ almost all instances in the phase-transition region are hard (see last column of Table 8.1). Also, as Table 8.1 shows, the number of hard instances varies a lot for the different heuristics. Therefore, it is not possible to compare the percentile running times of the different heuristics for $n > 40$. For $n = 80$ and $d = 14$ (see last column of Table 8.1), for instance, the 50% and 99% percentile element of the $\mathcal{C}_8/dynamic/global$ heuristic is element No.36 and element No.72, while it is element No.141 and element No.280 of the $\widehat{\mathcal{H}}_8/dynamic/local$ heuristic, respectively.

For this reason we show the results only up to a size of $n = 40$ (see Figure 8.7). Again, we took the average of the different degrees from $d = 10$ to $d = 15$ in order to cover the whole phase-transition region. The order of the run times is the same for different combinations of computing constrainedness: $\mathcal{B} \gg \widehat{\mathcal{B}}, \mathcal{C}_8 \gg \mathcal{Q}_8, \widehat{\mathcal{H}}_8$, while $\widehat{\mathcal{H}}_8$ is in most cases the fastest. As for the $A(n,d,4.0)$ instances, the run times for $dynamic/global$ were much longer than the other combinations. The 99% percentile run times of the $static/global$ combination and for $\widehat{\mathcal{H}}_8$ and $\mathcal{Q}_8$ of the $dynamic/local$ combination are faster

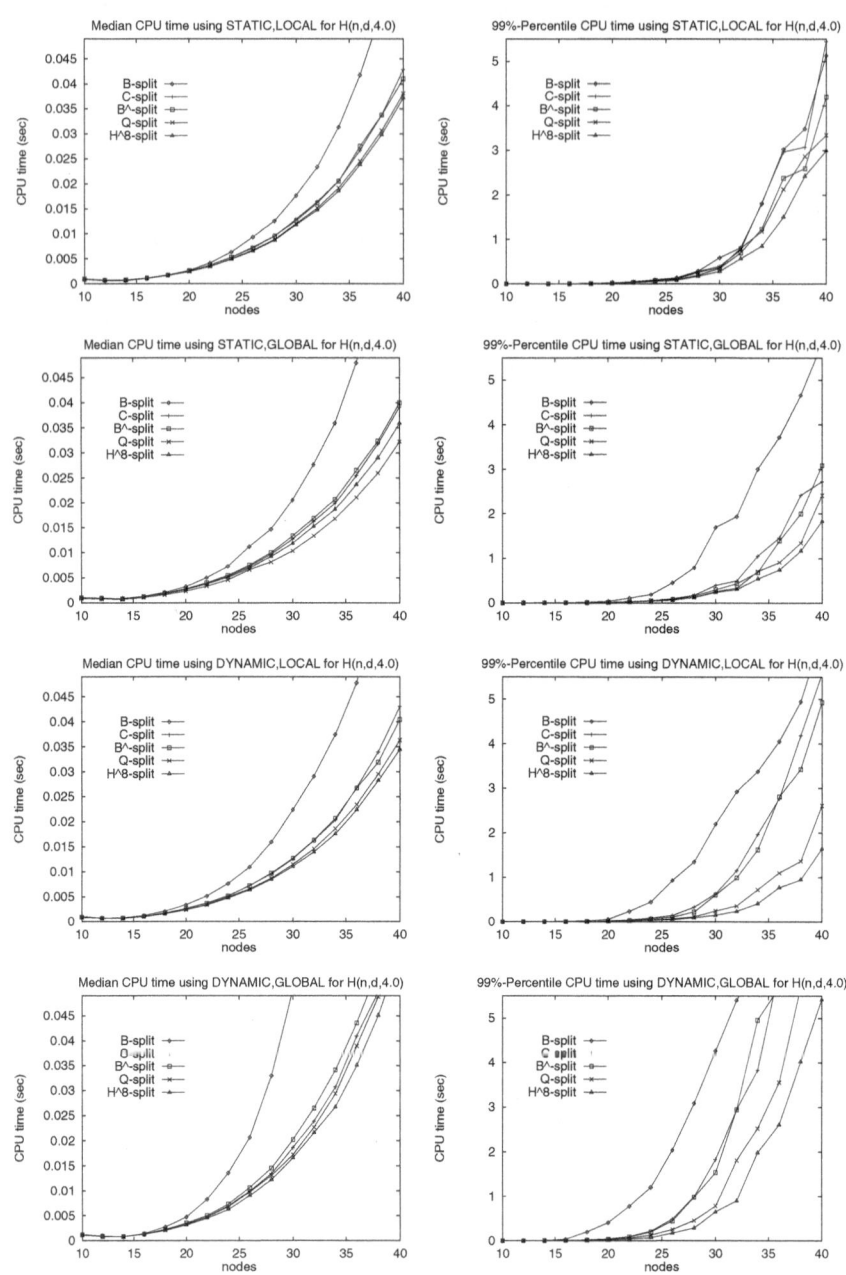

Fig. 8.7. Percentile 50% and 99% CPU time of the different heuristics for solving $H(n, d, 4.0)$ ($d = 10.0$ to $d = 15.0$, 5,500 instances per data point)

than those of the other combinations. Although the median CPU times are about the same as for $A(n, d, 4.0)$ for $n < 40$, the percentile 99% CPU times are much longer. As it was already shown in Figure 8.5 and Figure 8.4, this is a further evidence that there are very hard instances in the phase-transition region of $H(n, d, 4.0)$.

8.5 Orthogonal Combination of the Heuristics

In the previous section we studied the quality of different heuristics for solving randomly generated RSAT instances. We found that several instances which are mainly located in the phase-transition region could not be solved by some heuristics within our limit of 10,000 visited nodes in the search space. Since the different heuristics have a different search space (depending on the split-set) and use a different path through the search space (determined by the different possibilities of computing constrainedness), it is possible that instances are hard for some heuristics but easily solvable for other heuristics. Nebel [127] observed that running different heuristics in parallel can solve more instances of a particular hard set of temporal reasoning instances proposed by van Beek and Manchak [157] than any single heuristic alone can solve, when using altogether the same number of visited nodes than for each heuristic alone. An open question of Nebel's investigation [127] was whether this is also the case for the hard instances in the phase-transition region.

In this section we evaluate the power of "orthogonally combining" the different heuristics for solving RSAT instances, i.e., running the different heuristics for each instance in parallel until one of the heuristics solves the instance. There are different ways for simulating this parallel processing on a single processor machine. One is to use time slicing between the different heuristics, another is to run the heuristics in a fixed or random order until a certain number of nodes in the search space is visited and if unsuccessful try the next heuristic (cf. [92].) Which possibility is chosen and with which parameters (e.g., the order in which the heuristics are run and the number of visited nodes which is spent for each heuristic) determines the efficiency of the single processor simulation of the orthogonal combination. In order to find the best parameters, we ran all heuristics using at most 10,000 visited nodes for each heuristic on the set of hard instances identified in the previous section (those instances for which at least one heuristic required more than 10,000 visited nodes) and compared their behavior. Since we ran all heuristics on all instances already for the experiments of the previous section, we only had to evaluate their outcomes. This led to a very surprising result for the $A(n, d, 4.0)$ instances, namely, all of the 788 hard instances except for a single one were solved by at least one of the heuristics using less than 10,000 visited nodes. In Table 8.2 we list the percentage of hard instances that could be solved by the different heuristics and the percentage of first response by each of them when running the heuristics in parallel (i.e., which heuristic required

Table 8.2. Percentage of solved hard instances for each heuristic and percentage of first response when orthogonally running all heuristics. Note that sometimes different heuristics are equally fast. Therefore the sum is more than 100%

Heuristics	$A(n, d, 4.0)$		$H(n, d, 4.0)$	
	Solved Instances	1. Response	Solved Instances	1. Response
$\widehat{\mathcal{H}}_8$/sta/loc	91.88%	19.80%	71.86%	6.92%
$\widehat{\mathcal{H}}_8$/sta/glo	94.67%	12.56%	85.58%	14.26%
$\widehat{\mathcal{H}}_8$/dyn/loc	93.40%	24.37%	86.73%	22.28%
$\widehat{\mathcal{H}}_8$/dyn/glo	87.31%	13.58%	67.98%	15.00%
$\mathcal{C}_8$/sta/loc	89.72%	6.35%	61.60%	1.47%
$\mathcal{C}_8$/sta/glo	92.64%	5.20%	79.41%	5.04%
$\mathcal{C}_8$/dyn/loc	90.10%	5.96%	56.15%	2.26%
$\mathcal{C}_8$/dyn/glo	86.63%	6.60%	44.64%	2.40%
$\mathcal{Q}_8$/sta/loc	89.72%	9.77%	67.78%	1.63%
$\mathcal{Q}_8$/sta/glo	93.15%	12.06%	82.43%	3.61%
$\mathcal{Q}_8$/dyn/loc	90.61%	10.15%	81.72%	1.83%
$\mathcal{Q}_8$/dyn/glo	86.80%	12.82%	60.78%	4.61%
$\widehat{\mathcal{B}}$/sta/loc	91.37%	1.40%	68.42%	1.84%
$\widehat{\mathcal{B}}$/sta/glo	88.71%	1.27%	81.58%	5.22%
$\widehat{\mathcal{B}}$/dyn/loc	91.12%	0.89%	60.32%	2.56%
$\widehat{\mathcal{B}}$/dyn/glo	79.44%	0.89%	–	1.83%
$\mathcal{B}$/sta/loc	79.31%	0.51%	–	1.67%
$\mathcal{B}$/sta/glo	71.83%	0.25%	–	1.13%
$\mathcal{B}$/dyn/loc	73.48%	0.51%	–	0.42%
$\mathcal{B}$/dyn/glo	–	0.13%	–	0.49%
combined	99.87%		96.48%	

the smallest number of visited nodes for solving the instance). It turns out that the heuristics using $\widehat{\mathcal{H}}_8$ as a split-set did not only solve more instances than the other heuristics, they were also more often the fastest in finding a solution. Although the heuristics using the other two maximal tractable subsets $\mathcal{Q}_8$ and $\mathcal{C}_8$ as a split-set did not solve significantly more instances than the heuristics using $\widehat{\mathcal{B}}$, they were much faster in finding a solution. Despite solving the least number of instances, the heuristics using $\mathcal{B}$ as a split-set were in some cases the fastest in producing a solution.

When comparing the minimal number of visited nodes of all the heuristics for all the hard instances, we found that only five of them (which were all inconsistent) required more than 150 visited nodes. This is particularly remarkable as all these instances are from the phase-transition region of an NP-hard problem, i.e., instances which are usually considered to be the most difficult ones. Further note that about 15%(120) of the 788 (path-consistent) instances were inconsistent, which is much higher than usual (cf. Figure 8.4). Interestingly, most of those inconsistent instances were solved faster than the consistent instances. At this point, it should be noted that combining heuristics orthogonally is very similar to randomized search techniques with restarts [150]. However, in contrast to randomized search, our method can also determine whether an instance is inconsistent. In Figure 8.8 we chart the number of hard instances solved with the smallest number of visited nodes with respect to their solubility. Due to the low number of hard instances of $A(n, d, 4.0)$, the figure on the left looks a bit ugly but one can at least approximate the behavior of the curves when comparing it with the second figure on the right which is the same curve for $H(n, d, 4.0)$ (see below.) The oscillating behavior of the inconsistent instances (more instances are solved with an odd than with an even number of visited nodes) might be due to the sizes of the instances—we generated only instances with an even number of nodes. The most difficult instance $(n = 56, d = 10)$ was solved as inconsistent

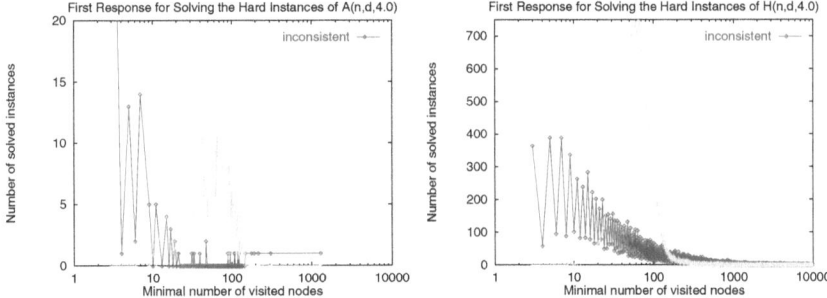

Fig. 8.8. Fastest solution of the hard instances when running all heuristics in parallel

with the $\widehat{\mathcal{B}}/static/global$ heuristic using about 91,000 visited nodes while all heuristics using one of the maximal tractable subsets as a split-set failed to solve it even when each was allowed to visit 20,000,000 nodes in the search space.

We did the same examination for the set of 75,081 hard instances of $H(n, d, 4.0)$. 2,640 of these instances could not be solved by any of the 20 different heuristics using 10,000 visited nodes each. Their distribution is shown in Figure 8.9(a). Similar to the hard instances of $A(n, d, 4.0)$, the heuristics using $\mathcal{H}_8$ as a split-set were the most successful ones for solving the hard instances of $H(n, d, 4.0)$, as shown in Table 8.2. They solved more of the hard instances than any other heuristics and produced the fastest response of more than 50% of the hard instances. There is no significant difference between using $\mathcal{C}_8, \mathcal{Q}_8$, or $\widehat{\mathcal{B}}$ as a split-set, neither in the number of solved instances nor in the percentage of first response. Like in the previous case, computing constrainedness using the *static/global* or the *dynamic/local* heuristics resulted in more successful paths through the search space by which more instances were solved within 10,000 visited nodes than by the other combinations. Also they produced on average faster solutions than the other combinations.

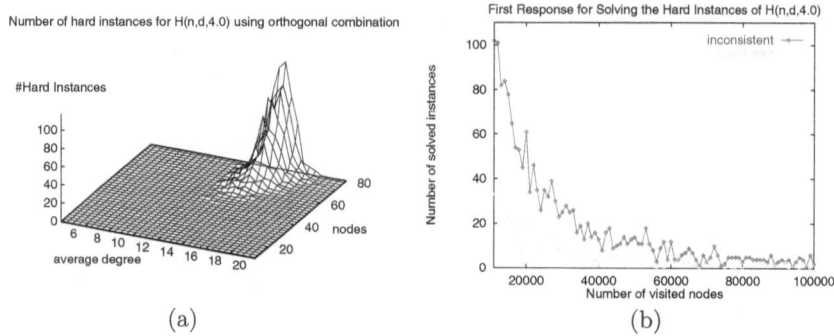

Fig. 8.9. Hard instances using orthogonal combination of all heuristic for $H(n, d, 4.0)$, (a) shows their distribution, (b) shows their fastest solution when using up to 100,000 visited nodes per heuristic

The same observations as for $A(n, d, 4.0)$ can be made when charting the fastest solutions of the hard instances of $H(n, d, 4.0)$ (see Figure 8.8). About 29% (21,307) of the solved instances are inconsistent. Most of them were, again, solved faster than the consistent instances. More than 75% of the hard instances can be solved with at most 150 visited nodes. 90% can be solved with at most 1,300 visited nodes. Since the $\widehat{\mathcal{H}}_8/dynamic/local$ heuristic alone solves more than 86% of the instances, it seems difficult to combine different heuristics in a way that more hard instances can be solved while using together not more than 10,000 visited nodes. However, when orthogonally combining the two best performing heuristics ($\widehat{\mathcal{H}}_8/dynamic/local$ and $\widehat{\mathcal{H}}_8/static/global$) allowing each of them a maximal number of 5,000 visitable nodes, we can solve 92% (69,056) of the hard instances.

We tried to solve the 2,640 hard instances of $H(n, d, 4.0)$ which are not solvable using orthogonal combination of heuristics with at most 10,000 visited nodes by using a maximal number of 100,000 visited nodes. 471 of these instances are still not solvable, more than 75% of the solved instances are inconsistent. The fastest response for the solved instances is charted in Figure 8.9(b). The most successful heuristics in giving the fastest response are $\widehat{\mathcal{H}}_8/dynamic/local$ (42.5%) and $\dot{\mathcal{H}}_8/static/global$ (26.6%). The three heuristics using $static/global$ computation of constrainedness combined with using $\mathcal{Q}_8, \mathcal{C}_8$, and $\widehat{\mathcal{B}}$ as a split-set gave the fastest response for 15.9% of the solved instances where the $\widehat{\mathcal{B}}$ strategy was by far the best among the three (9.4%).

8.6 Combining Heuristics for Solving Large Instances

In the previous section we found that combining different heuristics orthogonally can solve more instances using the same amount of visited nodes than any heuristic alone can solve. In this section we use these results in order to identify the size of randomly generated instances up to which almost all of them, especially those in the phase-transition region, can still be solved in acceptable time. Since many instances of $H(n, d, 4.0)$ are already too difficult for a size of $n = 80$ (see Figure 8.9), we restrict our analysis to the instances of $A(n, d, 4.0)$ and study randomly generated instances with a size of more than $n = 100$ nodes.

For instances of a large size allowing a maximal number of 10,000 visited nodes in the search space is too much for obtaining an acceptable runtime. 10,000 visited nodes for instances of size $n = 100$ corresponds to a runtime of more than 10 seconds on a Sun Ultra1, for larger instances it gets much slower. Therefore, we have to restrict the maximal number of visited nodes in order to achieve an acceptable runtime. Given a multi-processor machine, the different heuristics can each be run orthogonally on a different processor for the maximal number of visited nodes. If the orthogonal combination of the different heuristics is simulated on a single-processor machine, the maximal number of nodes has to be divided by the number of used heuristics to obtain the available number of visitable nodes for each heuristic. Thus, the more different heuristics we use, the less number of visitable nodes we can spend on each heuristic. Therefore, in order to achieve the best performance, we have to find the combination of heuristics that solves most instances within a given number of visitable nodes. The chosen heuristics should not only solve many instances alone, they should also complement each other well, i.e., instances which cannot be solved by one heuristic should be solvable by the other heuristic.

We started by finding the optimal combination of heuristics for the set of 788 hard instances of $A(n, d, 4.0)$. From our empirical evaluation given in Section 8.4 we know how many visited nodes each heuristic needs in order to solve each of the 788 hard instances. Therefore, we computed the number of solved instances for all 2^{20} possible combinations of the heuristics using an increasing maximal number of visitable nodes for all heuristics together. Since we only tried to find the combination which solves the most instances, this can be computed quite fast. The results are given in Table 8.3. They show that a good performance can be obtained with a maximal number of 600 visited nodes. In this case four heuristics were involved, i.e., 150 visitable nodes are spend for each of the four heuristics. Since the same combination of heuristics ($\widehat{\mathcal{H}}_8/static/global$, $\widehat{\mathcal{H}}_8/dynamic/local$, $\mathcal{C}_8/dynamic/local$, $\widehat{\mathcal{B}}/static/local$) is also the best for up to 1,000 visitable nodes, we choose this combination for our further analysis. We choose the order in which they are processed to be 1. $\widehat{\mathcal{H}}_8/dynamic/local$, 2. $\widehat{\mathcal{H}}_8/static/global$, 3. $\mathcal{C}_8/dynamic/local$, 4. $\widehat{\mathcal{B}}/static/local$ according to their first response behavior given in Table 8.2. Note that although the two heuristics $\mathcal{C}_8/dynamic/local$ and $\widehat{\mathcal{B}}/static/local$ do not show a particularly good performance when running them alone (see Table 8.2), they seem to best complement the other two heuristics.

Table 8.3. Best performance of combining different heuristics for solving the 787 solvable hard instances of $A(n, d, 4.0)$ with a fixed maximal number of visited nodes

Max Nodes	Solved Instances	Combination of Heuristics
100	516	$\widehat{\mathcal{H}}_8$-d-l
200	705	$\widehat{\mathcal{H}}_8$-s-g
300	759	$\widehat{\mathcal{H}}_8$-s-g, $\widehat{\mathcal{H}}_8$-d-l
400	769	$\widehat{\mathcal{H}}_8$-s-g, $\mathcal{C}_8$-d-l
500	774	$\widehat{\mathcal{H}}_8$-s-g, $\widehat{\mathcal{H}}_8$-d-l, $\mathcal{C}_8$-d-l
600	778	$\widehat{\mathcal{H}}_8$-s-g, $\widehat{\mathcal{H}}_8$-d-l, $\mathcal{C}_8$-d-l, $\widehat{\mathcal{B}}$-s-l
700	780	$\widehat{\mathcal{H}}_8$-s-g, $\widehat{\mathcal{H}}_8$-d-l, $\mathcal{C}_8$-d-l, $\widehat{\mathcal{B}}$-s-l
800	783	$\widehat{\mathcal{H}}_8$-s-g, $\widehat{\mathcal{H}}_8$-d-l, $\mathcal{C}_8$-d-l, $\widehat{\mathcal{B}}$-s-l
900	784	$\widehat{\mathcal{H}}_8$-s-g, $\widehat{\mathcal{H}}_8$-d-l, $\mathcal{C}_8$-d-l, $\widehat{\mathcal{B}}$-s-l
1100	785	$\widehat{\mathcal{H}}_8$-s-g, $\widehat{\mathcal{H}}_8$-d-l, $\mathcal{C}_8$-d-l, $\widehat{\mathcal{B}}$-s-l, $\widehat{\mathcal{B}}$-s-g
1300	786	$\widehat{\mathcal{H}}_8$-s-g, $\widehat{\mathcal{H}}_8$-d-l, $\widehat{\mathcal{B}}$-s-l, $\widehat{\mathcal{B}}$-s-g
3900	787	$\widehat{\mathcal{H}}_8$-s-g, $\widehat{\mathcal{H}}_8$-d-l, $\widehat{\mathcal{B}}$-d-l

What we have to find next is the maximal number of visitable nodes we spend for the heuristics. For this we ran the best performing heuristic ($\widehat{\mathcal{H}}_8/dynamic/local$) on instances of the phase-transition region of varying sizes. It turned out that for almost all consistent instances the number of visited nodes required for solving them was slightly less than twice the size of the instances while most inconsistent instances are also not path-consistent and, thus, solvable with only one visited node. Therefore, we ran the four heuristics in the following allowing $2n$ visited nodes each, where n is the size of the instance, i.e., together we allow at most $8n$ visitable nodes. We randomly generated test instances according to the $A(n, d, 4.0)$ model for a size of $n = 110$ regions up to a size of $n = 500$ regions with a step of 10 regions and 100 instances for each size and each average degree ranging from $d = 2.0$ to $d = 18.0$ with a step of 0.5, a total number of 132,000 instances. Since solving large instances using backtracking requires a lot of memory, we solved the instances on a Sun Ultra60 with 1GB of main memory.

The generated instances display a phase-transition behavior which continues the one given in Figure 8.2. The phase-transition ranges from $d = 10.0$ for $n = 110$ to $d = 10.5$ for $n = 500$ (see Figure 8.10). Apart from 112 instances, all other instances we generated were solvable by orthogonal combination of the four heuristics ($\widehat{\mathcal{H}}_8/static/global$, $\widehat{\mathcal{H}}_8/dynamic/local$, $\mathcal{C}_8/dynamic/local$, $\widehat{\mathcal{B}}/static/local$) spending less than 2n visited nodes each. In Figure 8.10 we give the average number of visited nodes of the path-consistent instances. It can be seen that for our test instances the average number of visited nodes is linear in the size of the instances. The percentile 70% CPU time for instances of the phase-transition with a size of $n = 500$ regions is about

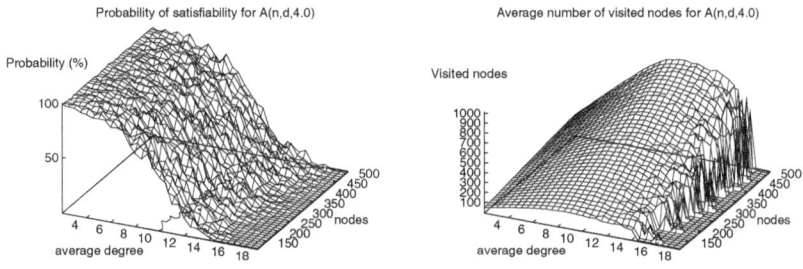

Fig. 8.10. Probability of satisfiability for $A(n, d, 4.0)$ (100 instances per data point) and average number of visited nodes of the path-consistent instances when using orthogonal combination of the four selected heuristics

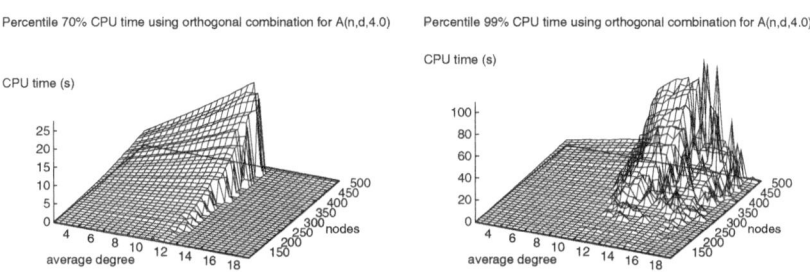

Fig. 8.11. Percentile 70% and 99% CPU time of the orthogonal combination of four different heuristics for solving large randomly generated instances of $A(n, d, 4.0)$

20s, the percentile 99% CPU time is about 90s. Up to a size of $n = 400$ regions, the percentile 99% CPU time is less than a minute (see Figure 8.11).

131,240 of our test instances were already solved by the $\widehat{\mathcal{H}}_8/static/global$ heuristic, for 71 instances the $\widehat{\mathcal{H}}_8/dynamic/local$ heuristic was required and for 577 instances the $\mathcal{C}_8/dynamic/local$ heuristic. None of the 112 instances which were not solved by one of the previous three heuristics were solve by the $\widehat{\mathcal{B}}/static/local$ heuristic. We tried to solve these instances using the other heuristics, again using a maximal number of $2n$ visited nodes each. The best performing among those heuristics was the $\mathcal{C}_8/dynamic/global$ heuristic which solved 87 of the 112 instances followed by the $\mathcal{C}_8/static/global$ heuristic (83) and the $\mathcal{Q}_8/dynamic/global$ heuristic (63). 7 instances were not solved by any heuristic within a maximal number of $2n$ visited nodes.

8.7 Discussion

We empirically studied the behavior of solving randomly generated RSAT instances using different backtracking heuristics some of which make use of the maximal tractable subsets identified in previous work. We generated instances according to two different models of which the "general model" A allows all 256 RCC-8 relations to be used while the "hard model" H allows only relations which are not contained in any of the maximal tractable subsets. A theoretical analysis of the two models showed that the model H and the model A for a small average degree do not suffer from trivial local inconsistencies as it is the case for similar generation procedures for CSPs with finite domains (cf. [2].) It turned out that randomly generated instances of both models show a phase-transition behavior which depends most strongly on the average degree of the instances. While most instances outside the phase-transition region can be solved efficiently by each of our heuristics, instances in the phase-transition region can be extremely hard. For the instances of the general model, most path-consistent instances are also consistent. Conversely, path-consistency is a very bad approximation to consistency for instances of the hard model. These instances are also much harder to solve than instances of the general model.

When comparing the different heuristics, we found that the heuristics using one of the maximal tractable subsets as a split-set are not as much faster in deciding consistency of RSAT instances as their theoretical advantage given by the reduced average branching factor and the resulting exponentially smaller size of the search space indicates. This is because using path-consistency as a forward checking method reduces the search space in all cases considerably. Nevertheless, using one of the maximal tractable subsets as a split-set, in particular $\widehat{\mathcal{H}}_8$, still leads to a much faster, though not exponentially faster, solution and solves more instances in reasonable time than the other heuristics. Although the two maximal tractable subsets $\mathcal{Q}_8$ and $\mathcal{C}_8$ contain more relations than $\widehat{\mathcal{H}}_8$ their average branching factor is lower, i.e., when using $\widehat{\mathcal{H}}_8$ one has to decompose more relations ($256 - 148 = 108$) than when using the other two sets (96, resp. 98 relations), but $\widehat{\mathcal{H}}_8$ splits the relations better than the other two sets. Most relations can be decomposed into only two $\widehat{\mathcal{H}}_8$ sub-relations, while many relations must be decomposed into three $\mathcal{C}_8$ or into three $\mathcal{Q}_8$ sub-relations. This explains the superior performance of heuristics involving $\widehat{\mathcal{H}}_8$ for decomposition.

We stored all instances which could not be solved by all heuristics within a maximum number of 10,000 visited nodes in the search space in order to find out how the different heuristics perform on these hard instances. We found that almost all hard instances are located in the phase-transition region and that there are many more hard instances in the hard model than in the general model. We combined all heuristics orthogonally and ran them on all hard instances. This turned out to be very successful. Apart for one instance, all hard instances of the general model could be solved, most of them with a very low number of visited nodes. The hard instances of the hard model were much more difficult: many of them could not be solved by any of the heuristics. Nevertheless, many more instances were solved by orthogonally combining the heuristics than by each heuristic alone. Again, most of them were solved using a low number of visited nodes.

Based on our observations on orthogonally combining different heuristics, we tried to identify the combination of heuristics which is most successful in efficiently solving many instances and used this combination for solving very large instances. It turned out that the best combination involves only heuristics which use maximal tractable subsets for decomposition. With this combination we were able to solve almost all randomly generated instances of the phase-transition region of the general model up to a size of $n = 500$ regions very efficiently. This seems to be impossible when considering the enormous size of the search space, which is on average 10^{39323} for instances of size $n = 500$ when using $\widehat{\mathcal{H}}_8$ as a split-set.

Our results show that despite its NP-hardness, we were able to solve almost all randomly generated RSAT instances of the general model efficiently. This is neither due to the low number of different RCC-8 relations (instances generated according to the hard model are very hard in the phase-transition region) nor to our generation procedure for random instances which does not lead to trivially flawed instances asymptotically. It is mainly due to the maximal tractable subsets which cover a large fraction of RCC-8 and which lead to extremely low branching factors. Since there are different maximal tractable subsets, they allow one to choose between many different backtracking heuristics which further increases efficiency: some instances can be solved easily by one heuristic, other instances by other heuristics. Heuristics involving maximal tractable subclasses showed the best behavior but some instances can be solved faster when other tractable subsets are used. The full classification of tractable subsets gives the possibility of generating hard instances with a high probability. Many randomly generated instances of the phase-transition region are very hard when using only relations which are not contained in any of the tractable subsets and an instance size of more than $n = 60$ regions. The next step in developing efficient reasoning methods for RCC-8 is to find methods which are also successful in solving most of the hard instances of the hard model.

The results of our empirical evaluation of reasoning with RCC-8 suggest that analyzing the computational properties of a reasoning problem and identifying tractable subclasses of the problem is an excellent way for achieving efficient reasoning mechanisms. In particular maximal tractable subclasses can be used to develop more efficient methods for solving the full problem since their average branching factor is the lowest. Using the refinement method developed in Chapter 7, tractable subclasses of a set of relations forming a relation algebra can be identified almost automatically. This method makes it very easy to develop efficient algorithms. A further indication of our empirical evaluation is that it can be much more effective (even and especially for hard instances of the phase-transition region) to orthogonally combine different heuristics than to try to get the final epsilon out of a single heuristic. This answers a question raised by Nebel [127] of whether the orthogonal combination of heuristics is also useful in the phase-transition region. In our experiments this lead to much better results even when simulating the orthogonal combination of different heuristics on a single processor machine and spending altogether the same resources as for any one heuristic alone. In contrast to the method of time slicing between different heuristics, we started a new heuristic only if the previous heuristic failed after a certain number of visited nodes in the search space. The order in which we ran the heuristics depended on their performance and on how well they complemented each other, more successful heuristics were used first. This is similar to using algorithm portfolios as proposed by Huberman et al. [92]. Which heuristics perform better and which combination is the most successful one is a matter of empirical evaluation and depends on the particular problem. Heuristics depending on maximal tractable subclasses, however, should lead to the best performance.

For CSPs with finite domains there are many theoretical results about localizing the phase-transition behavior and about predicting where hard instances are located. In contrast to this, there are basically no such theoretical results for CSPs with infinite domains as used in spatial and temporal reasoning. As our initial theoretical analysis shows, theoretical results on CSPs with finite domains do not necessarily hold for CSPs with infinite domains as well. It would be very interesting to develop a more general theory for CSPs with infinite domains, possibly similar to Williams and Hogg's "Deep Structure" [167] or Gent et al.'s "Kappa" theory [65].

9. Representational Properties of RCC-8

The topological relationships distinguished by the Region Connection Calculus RCC-8 are very basic. Their usefulness is shown by the independent development of these relationships by Egenhofer [44] in the area of geographic information systems and by the existing and possible applications. As we have seen in Chapter 5, these relationships are also very interesting from a cognitive point of view.

Equally important as the question of whether a system of relations is useful is the question of what entities shall be described by the relations, i.e., how useful is the domain of the relations. For RCC-8, spatial regions are non-empty regular subsets of a topological space. This notion of spatial regions is very general, and the obvious question is whether this notion is too general so that it is maybe not useful for spatial reasoning. For instance, a topological space does not have a particular dimension, whereas most applications of qualitative spatial reasoning deal only with two- or three-dimensional space. It might, thus, be possible that a set of RCC-8 constraints is consistent but not realizable within a particular dimension. Lemon [114] gave an example of a set of RCC-8 constraints which is realizable in three dimensional space but not in two dimensional space if regions must be internally connected. Lemon used this result to argue that spatial logics like RCC are not an adequate formalism for representing space.

In addition to analyzing the computational properties of reasoning over RCC-8 relations, we must also clarify the connection between consistency of a set of RCC-8 constraints and its realizability within a particular dimension. If there were consistent sets of RCC-8 relations that are not realizable in the desired dimension, the domain of spatial regions of RCC-8 would be too general and would have to be restricted in order to be useful for spatial reasoning. It appears that this question also depends on the internal connectedness of regions. Although properties of internal connectedness such as the number of different pieces of a region cannot be discriminated by the RCC-8 relations, we will take this factor into account.

Apart from just knowing whether a consistent set of RCC-8 constraints is realizable it is also important to find a realization. This depends on the ability to represent spatial regions. Within Allen's interval algebra [4], for example, intervals can be represented as a pair of two points, the start and

J. Renz: Qualitative Spatial Reasoning with Topological Information, LNAI 2293, pp. 155–172, 2002.
© Springer-Verlag Berlin Heidelberg 2002

the end point of the intervals. Such a simple representation is not possible for spatial regions of higher dimensions as used by RCC-8. These regions might be arbitrary subsets of a topological space which are not necessarily analytically describable and which might consist of a large number of pieces. Therefore, it appears to be difficult to represent spatial regions as used by RCC-8 within a computational framework.

In this chapter we will refer to these representational topics. In order to represent arbitrary spatial regions, it is necessary to have a *canonical model* of RCC-8, i.e., a structure that allows to model any consistent sentence of the calculus. Topological space is of course a canonical model, but, as described above, this does not seem to be very useful for representing regions. Instead, we develop a canonical model of our modified version of the modal encoding of RCC-8 as introduced by Bennett [12]. In order to use this model for representing topological regions and to derive topological properties of regions, we have to transform this modal canonical model to a topological canonical model. This new canonical model of RCC-8 permits a simple representation of spatial regions by reducing them to their necessary topological features with respect to their spatial relations.

By using the canonical model, we will prove that for any consistent set of RCC-8 constraints there are realizations in any dimension $d \geq 1$ when regions are not forced to be internally connected. This is still true even when regions are constrained to be sets of polytopes. Actually, internal connectedness of regions is not at all forced in the RCC-theory, so RCC can still be seen as an adequate representation formalism of space. We will also argue that forcing internal connectedness of all regions is too restrictive when dealing with spatial regions. Nevertheless, we will prove that in three- and higher dimensional space every consistent set of RCC-8 constraints can always be realized with internally connected regions. Using the new canonical model for representing spatial regions, it becomes possible to determine realizations of consistent sets of RCC-8 constraints. We will give algorithms for generating realizations of both internally connected and disconnected regions.

9.1 A Canonical Model of RCC-8

In this section we identify a canonical model of RCC-8 which is based on the modal encoding of RCC-8 as given in Section 4.3. Modal formulas can be modeled by Kripke models, so our goal is to find a simple Kripke frame structure which allows to build a Kripke model for every consistent set of RCC-8 constraints. As we have already shown in Section 6.2, there is a particular Kripke model, the RCC-8 model, for the modal encoding $m(\Theta)$ of every consistent set of RCC-8 constraints Θ. Based on this particular RCC-8-model, we can now define a canonical model for RCC-8 (cf. Definition 6.14).

Definition 9.1 (RCC-8-structure, RCC-8-cluster).
An RCC-8-structure of size n $\mathcal{S}_{RCC8}^n = \langle W, \{R_\square, R_{\mathbf{I}}\}, \pi \rangle$ has the following properties:

1. *W contains only worlds of level 0 and 1.*
2. *For every world u of level 0 there are exactly $2n$ worlds v of level 1 with $uR_{\mathbf{I}}v$. These $2n + 1$ worlds form an RCC-8-cluster of size $2n + 1$ (cf. Figure 9.1).*
3. *For every world v of level 1 there is exactly one world u of level 0 with $uR_{\mathbf{I}}v$.*
4. *For all worlds $w, v \in W_{\mathcal{S}}$: $wR_{\mathbf{I}}w$ and $wR_\square v$.*

$\mathcal{S}_{RCC8}^n$ *contains RCC-8-clusters of size $2n + 1$ with all possible valuations[1] with respect to $R_{\mathbf{I}}$.*

Every RCC-8 model $\mathcal{M}$ of $m(\Theta)$ consists of a polynomial number of RCC-8 clusters of size $2n + 1$, where n is the number of variables of Θ. Thus, $\mathcal{M}$ is a subset of $\mathcal{S}_{RCC8}^n$. In order to obtain a Kripke frame structure that allows a model of $m(\Theta)$ for any number of variables in Θ, we define the RCC-8-structure $\mathcal{S}_{RCC8}$ as the union of all RCC-8-structures $\mathcal{S}_{RCC8}^n$ of any size $n \geq 1$. Since for every consistent set of RCC-8 constraints Θ there is an RCC-8-model of $m(\Theta)$ based on $\mathcal{S}_{RCC8}$, the RCC-8-structure is a canonical model of RCC-8.

9.2 A Topological Interpretation of the Canonical Model

The modal encoding of RCC-8 was obtained by introducing a modal operator $\mathbf{I}$ corresponding to the topological interior operator and transferring the topological properties and axioms to modal logic. In this section we present a way of topologically interpreting RCC-8-models such that all parts of the models can be interpreted consistently on a topological level.

Because $\mathbf{I}$ is an S4-operator and because of the additional constraints $m_2(x)$, exactly one of the following formulas is true for every world w of $\mathcal{M}$ and every propositional atom X (see Section 6.2.1).

1. $\mathcal{M}, w \Vdash \mathbf{I}X$
2. $\mathcal{M}, w \Vdash \mathbf{I}\neg X$
3. $\mathcal{M}, w \Vdash X \wedge \neg \mathbf{I}X$

Consider a particular world w. Then the set of all spatial variables can be divided into three disjoint sets of spatial variables according to which of the three possible formulas is true in w (see Figure 9.1). Let $\mathcal{I}_w$, $\mathcal{E}_w$, and $\mathcal{B}_w$ be the sets of spatial variables where the first, the second, and the third formula

---

[1] As the number of spatial variables is countable, the number of RCC-8-clusters with different valuations is also countable.

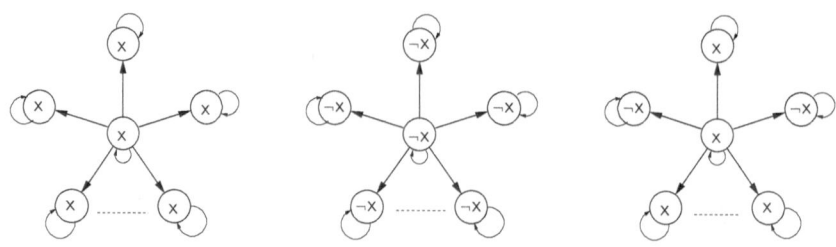

Fig. 9.1. Three possible RCC-8-clusters of the RCC-8-structure $\mathcal{S}_{RCC8}$

is true in w, respectively, i.e., $\mathcal{M}, w \Vdash \mathbf{IX} \wedge \mathbf{I}\neg\mathbf{Y} \wedge (\mathbf{Z} \wedge \neg\mathbf{IZ})$ for all $x \in \mathcal{I}_w$, $y \in \mathcal{E}_w$, and $z \in \mathcal{B}_w$.

When looking at points in a topological space, for every region there are three different kinds of points: interior points, exterior points, and boundary points of a region. If a point is an interior or exterior point of a region, there is a neighborhood of the point such that all points of the neighborhood are inside or outside the region, respectively. If a point is a boundary point of a region, every neighborhood contains points inside and points outside the region (see Definition 2.3).

There seems to be a correspondence between worlds and points of a topological space, and between the accessibility relation $R_{\mathbf{I}}$ and topological neighborhoods. In the following lemma we further investigate this correspondence by deriving topological constraints from the modal formulas.

Lemma 9.2. *Let x and y be two spatial variables of Θ. Depending on which sets $\mathcal{I}_w, \mathcal{E}_w$, or $\mathcal{B}_w$ they are contained in for a world w, the following relations between x and y are impossible. This has some topological consequences on possible instantiations $\mathbf{X}, \mathbf{Y}$:*

x	y	*Impossible Relations $R(x,y)$*	*Consequences*
$\mathcal{I}_w$	$\mathcal{I}_w$	DC, EC	$i(\mathbf{X}) \cap i(\mathbf{Y}) \neq \emptyset$
$\mathcal{I}_w$	$\mathcal{E}_w$	TPP, NTPP, EQ	$i(\mathbf{X}) \cap e(\mathbf{Y}) \neq \emptyset$
$\mathcal{I}_w$	$\mathcal{B}_w$	DC, EC, TPP, NTPP, EQ	$i(\mathbf{X}) \cap b(\mathbf{Y}) \neq \emptyset$
$\mathcal{E}_w$	$\mathcal{E}_w$		–
$\mathcal{E}_w$	$\mathcal{B}_w$	TPP^{-1}, NTPP^{-1}, EQ	$e(\mathbf{X}) \cap b(\mathbf{Y}) \neq \emptyset$
$\mathcal{B}_w$	$\mathcal{B}_w$	DC, NTPP, NTPP^{-1}	$b(\mathbf{X}) \cap b(\mathbf{Y}) \neq \emptyset^2$

Proof. Most entries in the table follow immediately from the encoding of the relations in modal logic. The only more difficult entry is the relation $\mathsf{EC}(x,y)$ in the third line of the table. This relation is not possible because of the property $\Box(\mathbf{Y} \to \neg\mathbf{I}\neg\mathbf{IY})$ which states that for any world w that satisfies $\mathbf{Y}$

---

[2] If $\mathsf{PO}(x,y)$ holds, $\mathbf{X}$ and $\mathbf{Y}$ do not necessarily have a common boundary point if one of them is not internally connected. However, assuming $b(\mathbf{X}) \cap b(\mathbf{Y}) \neq \emptyset$ in this case does not contradict any RCC-8 constraint, since RCC-8 is not expressive enough to distinguish different kinds of partial overlap.

there is a world v with $wR_\mathbf{I}v$ that satisfies **IY**. As v also satisfies **IX**, the model constraint of $\mathsf{EC}(x, y)$ is violated, so this relation is not possible. The topological consequences result from distinguishing the impossible from the possible relations.   ∎

It can be seen that when, e.g., **IX** and **IY** hold in a world w, then X and Y must have a common interior. So, there is a common interior point of X and Y where w can be mapped to. In the following theorem we give a mapping of every world to a point in the topological space.

Theorem 9.3. *Let Θ be a consistent set of* RCC-8 *constraints, $m(\Theta)$ be the modal encoding of Θ, $\mathcal{M} = \langle W, \{R_\square, R_\mathbf{I}\}, \pi \rangle$ be an* RCC-8-*model of $m(\Theta)$, and $\mathcal{U}$ a topological space. Then there is a function $p : W \mapsto \mathcal{U}$ that maps each world $w \in W$ to a point $p(w) \in \mathcal{U}$ and a function $N : W \mapsto 2^\mathcal{U}$ that assigns each world $w \in W$ a neighborhood $N(w)$ of $p(w)$ such that $p(w)$ is in the interior of X if $\mathcal{M}, w \Vdash \mathbf{IX}$ holds, $p(w)$ is in the exterior of X if $\mathcal{M}, w \Vdash \mathbf{I}\neg\mathbf{X}$ holds, $p(w)$ is on the boundary of X if $\mathcal{M}, w \Vdash \mathbf{X}\wedge\neg\mathbf{IX}$ holds, and $p(u) \in N(w)$ if and only if $wR_\mathbf{I}u$ holds.*[3]

Proof. Let w be a world of W and $\mathcal{I}_w, \mathcal{E}_w$, and $\mathcal{B}_w$ be the corresponding sets of spatial variables. We assume that there is a realization of Θ such that there is at least one point in the topological space that is in the interior of every X, in the exterior of every Y, and on the boundary of every Z simultaneously ($x \in \mathcal{I}_w, y \in \mathcal{E}_w, z \in \mathcal{B}_w$). It follows from Lemma 9.2 that this is true for every pair of regions. As RCC-8 permits only binary constraints between spatial variables and regions are allowed to be internally disconnected, this assumption holds. We further assume that p maps w to one of these points.

Because of Definition 2.3, there must be neighborhoods $N_X(w)$ and $N_Y(w)$ of $p(w)$ for every $x \in \mathcal{I}_w$ and every $y \in \mathcal{E}_w$ such that $N_X(w)$ is in the interior of X and $N_Y(w)$ is disjoint with Y. Also, for every $z \in \mathcal{B}_w$, every neighborhood $N_Z(w)$ of $p(w)$ contains points inside and outside Z. All these neighborhoods are members of the neighborhood system of $p(w)$, so their intersection $N(w)$ is also a neighborhood of $p(w)$ where all $R_\mathbf{I}$-successors of w can be mapped to.   ∎

Using the above defined functions p and N, $\mathcal{M}, w \Vdash \mathbf{IX}$ can be interpreted as "there is a neighborhood $N(w)$ of $p(w)$ such that all points of $N(w)$ are in X". This obeys the intended interpretation of $\mathbf{I}$ as an interior operator, as $\mathcal{M}, w \Vdash \mathbf{X}$ means that $p(w)$ is in X and $\mathcal{M}, w \Vdash \mathbf{IX}$ means that $p(w)$ is in the interior of X.

The function N, as defined in Theorem 9.3, can be replaced by any function $N' : W \mapsto 2^\mathcal{U}$, with $N'(w) \subseteq N(w)$ for all $w \in W$, if $N'(w)$ is a member

---

[3] The properties for $R_\square$ ($p(u) \in \mathcal{U}$ if $wR_\square u$ holds and $p(w) \in \mathcal{U}$) can be omitted as we already defined N and p such that only points of $\mathcal{U}$ are used.

of the neighborhood system of $p(w)$. p has to be changed accordingly. In particular, we will regard in the following all neighborhoods as d-dimensional spheres where d is the dimension of the underlying topological space.

In order to make the following argumentation more readable, a world mapped to an interior point of X is denoted *interior world* of x, a world mapped to an exterior point of X *exterior world* of x, and a world mapped to a boundary point of X *boundary world* of x. Accordingly, a region is called *interior, exterior* or *boundary region* of a world. In particular, a world w with $\mathcal{M}, w \Vdash \mathsf{IX}$ is an interior world of x, a world w with $\mathcal{M}, w \Vdash \mathsf{I}\neg\mathsf{X}$ is an exterior world of x, and a world w with $\mathcal{M}, w \Vdash \mathsf{X} \wedge \neg\mathsf{IX}$ is a boundary world of x.

9.3 RCC-8 Models and the Dimension of Space

In the previous section we have shown how the RCC-8-models introduced in Section 9.1 can be mapped to topological space, but we still have no information about the dimension of the topological space. In this section we investigate the influence of dimension on the possibility to map the RCC-8-models to the topological space, i.e., which dimension is required in order to find a realization of a consistent set of RCC-8 constraints Θ. We will start with proving that for any RCC-8-model there is a realization in two-dimensional space. It is sufficient to prove this only for sets of base relations as every realization of Θ uses only base relations.[4] For this proof it is important to keep in mind that regions do not have to be internally connected, i.e., they might consist of different disconnected pieces. It will turn out that our proof leads to realizations in any dimension $d \geq 1$. Finally, for three- and higher-dimensional space we will prove that every consistent set Θ can also be realized with internally connected regions.

For the following analysis we restrict regions to be sets of d-dimensional polytopes. Sets are required since regions might consist of several disconnected pieces where each piece is a single polytope. This restriction will be lifted later and the results can be generalized to arbitrary regular regions.

Let Θ be a consistent set of RCC-8 constraints and $\mathcal{M}$ be an RCC-8-model of $m(\Theta)$, the modal encoding of Θ. Suppose that only two-dimensional regions are permitted, i.e., the topological space is a two-dimensional plane $\mathcal{U}$. All worlds of $\mathcal{M}$ are mapped to points of $\mathcal{U}$ as specified in Theorem 9.3. The general intuition of the proof is that every RCC-8-cluster, i.e., every world of level 0 together with its R_I-successors is mapped to an independent neighborhood such that each neighborhood can be placed on an arbitrary but distinct position on the plane. Each neighborhood will then be extended to different closed sets that form the pieces of the spatial regions. In the

---

[4] The relation EQ can be omitted as any pair of spatial variables x and y with $\mathsf{EQ}(x, y)$ can be combined to a single spatial variable.

following we will study the requirements neighborhoods have to meet in order to guarantee two-dimensional realizations.

For every spatial variable x_i $(1 \leq i \leq n)$ and every world w of level 0, we define a *region vector* $r_i^w = (r_{i,1}^w, \ldots, r_{i,2n}^w)$ that represents the affiliation of the $2n$ R_I-successors of w to X_i, i.e., $r_{i,j}^w = 1$ if $\mathcal{M}, v_j \Vdash X_i$ and $r_{i,j}^w = 0$ if $\mathcal{M}, v_j \nVdash X_i$ where v_j is the jth R_I-successor of w. Since in the two-dimensional case the neighborhood $N(w)$ is a circle, we suppose that the points $p(v_j)$ corresponding to the R_I-successors v_j of w are ordered clockwise within the circle according to j. If $p(w)$ is a boundary point of X_i, some values of r_i^w are 1 and some are 0. Otherwise all values of r_i^w are either 1 (if $p(w)$ is contained in X_i) or 0 (if $p(w)$ is not contained in X_i).

Lemma 9.4. *If for every world w of level 0 there is a permutation P_w of the values of r_i^w such that $(r_{i,P_w(1)}^w, \ldots, r_{i,P_w(2n)}^w)$ is a bitonic sequence[5] for all $1 \leq i \leq n$, then the neighborhoods $N(w)$ can be placed in a two-dimensional plane such that all spatial relations are satisfied within the neighborhoods.*

Proof. If r_i^w is a bitonic sequence and $p(w)$ is a boundary point of X_i, then the mappings of the worlds of level 1 corresponding to the values of r_i^w can be separated into points inside X_i and points outside X_i by at most two line segments meeting at $p(w)$ (see Figure 9.2). These line segments can

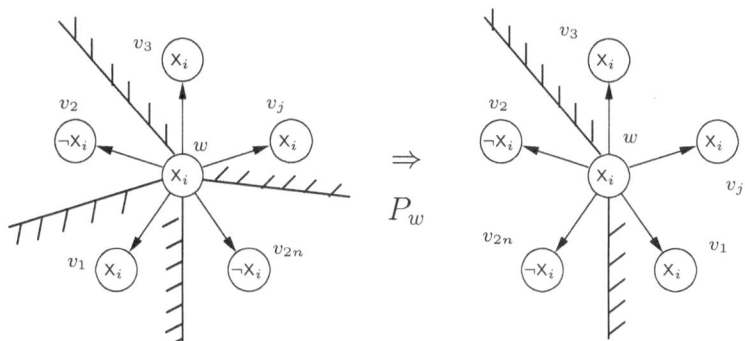

Fig. 9.2. Permutation P_w of the R_I-successors of a world w. The solid line indicates the boundary of X, the hashed region the interior of X

be regarded as the part of the boundary of X_i which is inside $N(w)$. So, neighborhoods can be separated in an interior and an exterior part of a region by a one-dimensional boundary. Therefore all neighborhoods can be placed in a two-dimensional plane. As the permutation of the R_I-successors has no influence on the relations between the regions, all spatial relations between the regions hold within the neighborhoods. ∎

---

[5] The values of a *bitonic* sequence are in a form $0^e 1^f 0^g$ or $1^e 0^f 1^g$ for $e, f, g \geq 0$ [32, p.642].

Actually, a permutation as described in the previous lemma is not necessary to guarantee two-dimensional realizations. A region might look as shown on the left of Figure 9.2, but in this case we restrict the shape and the internal connection of the regions by the neighborhoods we are using which is not at all desirable. However, a permutation as described in Lemma 9.4 is necessary for one-dimensional realizations and realizations with internally connected regions.

Since a permutation P_w is only necessary for boundary worlds, we will in the following try to keep the number of boundary worlds as small as possible. Therefore, we consider only RCC-8-models for which all boundary worlds are explicitly forced to be boundary worlds by the constraints. In order to do so, we have to take a closer look at which worlds are introduced as boundary worlds of some regions by the entailment constraints, and which worlds are forced to be boundary worlds of regions by the constraints. As a world w of level 0 is forced to be a boundary world of x if $\mathcal{M}, w \Vdash \mathsf{X}$ and $\mathcal{M}, v \not\Vdash \mathsf{X}$ hold for a world v with $wR_\mathbf{I}v$, we have to find out which of the model and entailment constraints force $\mathcal{M}, w \Vdash \mathsf{X}$ if $\mathcal{M}, v \not\Vdash \mathsf{X}$ holds or force $\mathcal{M}, v \not\Vdash \mathsf{X}$ if $\mathcal{M}, w \Vdash \mathsf{X}$ holds.

Proposition 9.5. *Boundary worlds are introduced only by the following relations (see Table 4.4):*

1. $\mathsf{EC}(x,y)$: $\neg\Box(\neg(\mathsf{X} \wedge \mathsf{Y}))$ *introduces a boundary world of x and y because of $\Box(\neg(\mathbf{IX} \wedge \mathbf{IY}))$.*
2. $\mathsf{TPP}(x,y)$: $\neg\Box(\mathsf{X} \rightarrow \mathbf{IY})$ *introduces a boundary world of x and y because of $\Box(\mathsf{X} \rightarrow \mathsf{Y})$.*
3. $\mathsf{TPP}^{-1}(x,y)$: $\neg\Box(\mathsf{Y} \rightarrow \mathbf{IX})$ *introduces a boundary world of x and y because of $\Box(\mathsf{Y} \rightarrow \mathsf{X})$.*

Apart from the above worlds that are introduced as boundary worlds of particular regions, worlds can also be forced to be boundary worlds of other regions.

Proposition 9.6. *A world w is forced to be a boundary world of x only with the following constraints:*

1. $\Box(\mathsf{X} \rightarrow \mathsf{Y})$: *If w is a boundary world of y and X is true in w, then w must also be a boundary world of x.*
2. $\Box(\mathsf{Y} \rightarrow \mathsf{X})$: *If w is a boundary world of y and $\neg\mathsf{X}$ is true in an $R_\mathbf{I}$-successor of w, then w must also be a boundary world of x.*
3. $\Box(\neg(\mathbf{IX} \wedge \mathbf{IY}))$: *If X and Y are true in w, then w must be a boundary world of x and y.*

For the constraints $\Box(\mathsf{X} \rightarrow \mathsf{Y})$ and $\Box(\mathsf{Y} \rightarrow \mathsf{X})$, w must already be a boundary world of some other region, so w must be introduced by one of the relations $\mathsf{EC}(x,y)$, $\mathsf{TPP}(x,y)$, or $\mathsf{TPP}^{-1}(x,y)$. If w is forced to be a boundary world of x and y with the constraint $\Box(\neg(\mathbf{IX} \wedge \mathbf{IY}))$, then X and Y must both be true in w. This can only be forced when there is a $z_1 \in Var(\Theta)$ with $\mathsf{TPP}(z_1, x)$ and

Z_1 is true in w, a $z_2 \in Var(\Theta)$ with $\mathsf{TPP}(z_2, y)$ and Z_2 is true in w, and w is a boundary world of z_1 and z_2 introduced by $\mathsf{EC}(z_1, z_2)$. So, in any case when a world is forced to be a boundary world of some region it must already be a boundary world of other regions introduced as described in Proposition 9.5.

We will now have a look at how regions must be related in order to force a world to be a boundary world of these regions using the constraints of Proposition 9.6. Suppose that w is a boundary world of x and y introduced by either $\mathsf{EC}(x, y)$ or $\mathsf{TPP}(x, y)$.[6] We will write $x|y$ in order to express that we can either use x or y but always the same. With one of the following constraints it can be forced that w is also a boundary world of $z \neq x, y$ (v is an R_I-successor of w):

$$\Box(Z \rightarrow (X|Y)) \text{ and } \mathcal{M}, w \Vdash Z \quad (\rightsquigarrow \mathsf{TPP}(z, x|y))$$
$$\Box(\neg(IZ \wedge I(X|Y))) \text{ and } \mathcal{M}, w \Vdash Z \quad (\rightsquigarrow \mathsf{EC}(z, x|y))$$
$$\Box((X|Y) \rightarrow Z) \text{ and } \mathcal{M}, v \Vdash \neg Z \quad (\rightsquigarrow \mathsf{TPP}^{-1}(z, x|y))$$

$\mathcal{M}, w \Vdash Z$ is forced with the following constraint:

$$\Box(U \rightarrow Z) \text{ and } \mathcal{M}, w \Vdash U \qquad (\rightsquigarrow \mathsf{TPP}(u, z))$$

$\mathcal{M}, v \Vdash \neg Z$ is forced with the following constraints:

$$\Box(Z \rightarrow U) \text{ and } \mathcal{M}, v \Vdash \neg U \qquad (\rightsquigarrow \mathsf{TPP}^{-1}(u, z))$$
$$\Box(\neg(IZ \wedge IU)) \text{ and } \mathcal{M}, v \Vdash U \qquad (\rightsquigarrow \mathsf{EC}(u, z))$$

When we compose these relations, we obtain the possible relations between u and $x|y$.

$R(u, z)$	$S(z, x|y)$	$(R \circ S)(u, x|y)$
TPP	TPP	TPP, NTPP
TPP	EC	DC, EC
TPP^{-1}	TPP^{-1}	TPP^{-1}, NTPP^{-1}
EC	TPP^{-1}	DC, EC

As w is a boundary world of x and y, $\mathsf{DC}(u, x|y)$ and $\mathsf{NTPP}(u, x|y)$ are not possible together with $\mathcal{M}, w \Vdash U$, and $\mathsf{NTPP}^{-1}(u, x|y)$ is not possible together with $\mathcal{M}, v \Vdash \neg U$. In order to force $\mathcal{M}, w \Vdash U$, there must be a sequence of spatial variables u_i with $\mathsf{TPP}(u_1, u)$, $\mathsf{TPP}(u_{i+1}, u_i)$, until there is a u_m that is equal to x or y, so $\mathsf{TPP}(x, z)$ or $\mathsf{TPP}(y, z)$ must hold. In order to force $\mathcal{M}, v \Vdash \neg U$, there must be a sequence of spatial variables u_i with $\mathsf{TPP}^{-1}(u_1, u)$, $\mathsf{TPP}^{-1}(u_{i+1}, u_i)$, and $\mathsf{EC}(u_j, u_{j-1})$ and $\mathcal{M}, v \Vdash U_j$ must hold. In order to force $\mathcal{M}, v \Vdash U_j$, there must be a sequence of TPP-related spatial variables, as described above, until one of them is equal to x or y, so $\mathsf{EC}(z, x)$ or $\mathsf{EC}(z, y)$ must hold. This results in only three different possibilities of how w is forced to be a boundary world of z if w was introduced as a boundary world of x and y.

---

[6] We omit $\mathsf{TPP}^{-1}(x, y)$ as $\mathsf{TPP}^{-1}(x, y) = \mathsf{TPP}(y, x)$ and the order is not important.

   a. $\mathsf{TPP}(x,y), \mathsf{TPP}(x,z)$, and $\mathsf{TPP}(z,y)$ hold.
   b. $\mathsf{EC}(x,y), \mathsf{TPP}(x,z)$, and $\mathsf{EC}(z,y)$ hold.
   c. $\mathsf{EC}(x,y), \mathsf{TPP}(y,z)$ and $\mathsf{EC}(z,x)$ hold.

As different spatial variables z_i, z_j, for which w is forced to be a bound-ary world of, all have the boundary world w in common, only the relations $\mathsf{EC}, \mathsf{PO}, \mathsf{TPP}$, or TPP^{-1} can hold between them.

   We have shown that only those worlds are boundary worlds which are introduced as boundary worlds of some regions by the entailment constraints, and, further, that other regions are only forced to be boundary regions of these worlds when they are related in a particular way. This will be used in the following lemma.

Lemma 9.7. *Let $\mathcal{M}$ be an* RCC-8-*model. Then two different types of $R_{\mathbf{I}}$-successors are sufficient for every world w of level 0.*

Proof. If w is not a boundary world of some region, all $R_{\mathbf{I}}$-successors of w satisfy exactly the same formulas as w. Otherwise, w is introduced as a boundary world by either $\mathsf{EC}(x,y)$ or $\mathsf{TPP}(x,y)$ (see Proposition 9.5). Let w be forced to be a boundary world of the spatial variables z_i. For $\mathsf{EC}(x,y)$, some of the $R_{\mathbf{I}}$-successors of w satisfy X but not Y, and some satisfy Y but not X, the others neither satisfy X nor Y. For all z_i with $\mathsf{TPP}(x, z_i)$ and $\mathsf{EC}(z_i, y)$ and all z_j with $\mathsf{TPP}(y, z_j)$ and $\mathsf{EC}(z_j, x)$, all $R_{\mathbf{I}}$-successors of w satisfy Z_i if they satisfy X and satisfy Z_j if they satisfy Y. So all $R_{\mathbf{I}}$-successors of w that satisfy X satisfy the same formulas, and all $R_{\mathbf{I}}$-successors of w that satisfy Y satisfy the same formulas. For the $R_{\mathbf{I}}$-successors v of w which do not satisfy X or Y, there are only two requirements: if $\mathsf{TPP}(z_{k'}, z_k)$ holds then Z_k must be true in v whenever $\mathsf{Z}_{k'}$ is true in v; if $\mathsf{EC}(z_k, z_{k'})$ holds then Z_k and $\mathsf{Z}_{k'}$ must not both be true in v. However, there is no constraint that forces the existence of these worlds, so it can be assumed that all $R_{\mathbf{I}}$-successors of w satisfy either X or Y. As the respective worlds all satisfy the same formulas, two different kinds of $R_{\mathbf{I}}$-successors of the boundary world w introduced by $\mathsf{EC}(x,y)$ are sufficient.

   For $\mathsf{TPP}(x,y)$, all $R_{\mathbf{I}}$-successors of w that satisfy X also satisfy Y, all $R_{\mathbf{I}}$-successors of w that do not satisfy Y also do not satisfy X, and some $R_{\mathbf{I}}$-successors of w satisfy Y but not X. For all z_i with $\mathsf{TPP}(x, z_i)$ and $\mathsf{TPP}(z_i, y)$, all $R_{\mathbf{I}}$-successors of w satisfy Z_i if they satisfy X. For the $R_{\mathbf{I}}$-successors of w that satisfy Y but not X, there is only one requirement, namely, that Z_k must be true whenever $\mathsf{Z}_{k'}$ is true in these worlds for any two spatial variables $z_k, z_{k'}$ with $\mathsf{TPP}(x, z_k), \mathsf{TPP}(z_{k'}, y)$ and $\mathsf{TPP}(z_k, z_{k'})$. However, there is again no constraint that forces the existence of these worlds, so it can be assumed that all $R_{\mathbf{I}}$-successors of w satisfy X if they satisfy Y. ∎

Whether a boundary world w is introduced by $\mathsf{EC}(x,y)$ or by $\mathsf{TPP}(x,y)$, in both cases two different kinds of $R_{\mathbf{I}}$-successors are sufficient. Thus, by grouping together the respective $R_{\mathbf{I}}$-successors for every world w of level 0 of

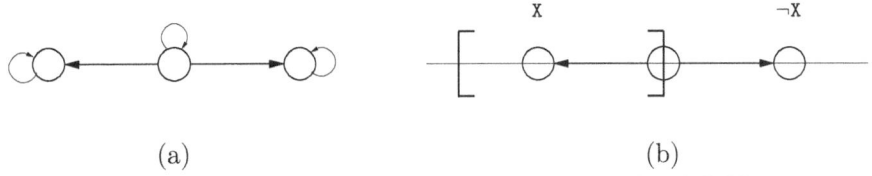

(a) (b)

Fig. 9.3. (a) shows a reduced RCC-8-cluster of the reduced RCC-8-structure. (b) shows how a neighborhood can be placed in one-dimensional space. The two brackets indicate a one-dimensional region X where the neighborhood represents a boundary point of X

$\mathcal{M}$, we can always find a permutation of the worlds of level 1 such that r^w is a bitonic sequence for all regions.

Instead of having $2n$ R_I-successors for every world of level 0 from which we know that they belong to only two different types, it is sufficient two use only two R_I-successors for every world of level 0. This leads to a very simple canonical model which is defined in the same way as in Definition 9.1 except that we have exactly two worlds of level 1 instead of $2n$ worlds.

Definition 9.8 (reduced RCC-8-structure/ -cluster/ -model).
A reduced RCC-8-structure $\mathcal{S}^*_{RCC8} = \langle W, \{R_\Box, R_I\}, \pi \rangle$ has the following properties:

1. *W contains only worlds of level 0 and 1.*
2. *For every world u of level 0 there are exactly two worlds v of level 1 with $uR_I v$. These three worlds form a reduced RCC-8-cluster (see Figure 9.3a).*
3. *For every world v of level 1 there is exactly one world u of level 0 with $uR_I v$.*
4. *For all worlds $w, v \in W_{\mathcal{S}}$: $wR_I w$ and $wR_\Box v$.*

$\mathcal{S}^*_{RCC8}$ *contains RCC-8-clusters with all possible valuations. The corresponding models are denoted as* reduced RCC-8-models.

It follows from Lemma 9.7 that the reduced RCC-8-structure is a canonical model for RCC-8. Note that the reduced RCC-8-structure is equivalent to an RCC-8-structure of size 1 and that a reduced RCC-8-cluster is equivalent to an RCC-8-cluster of size 3.

We can now apply Lemma 9.4 and place all neighborhoods independently on the plane while all relations between spatial regions hold within the neighborhoods. Thereby, neighborhoods corresponding to non-boundary worlds are homogeneous in the sense that all points within one of these neighborhoods have the same topological properties. Neighborhoods corresponding to a boundary world w consist of two homogeneous parts corresponding to the two R_I-successors of w. These two parts are divided by the common boundary of the boundary regions of w (see Figure 9.4a).

In order to obtain a realization, we have to find regions such that the relations between them hold in the whole plane and not just within the neighborhoods. Since regions do not have to be internally connected, it is possible to compose every region out of pieces resulting from the corresponding neighborhoods, i.e., for every neighborhood a region is affiliated with, we generate a piece of that region. As the neighborhoods are open sets and regions as well as their pieces must be regular closed sets, we have to *close* every neighborhood, i.e., find a closed set X^w for every region X and every neighborhood $N(w)$ with $\mathcal{M}, w \Vdash X$ such that all relations hold between the regions composed of the pieces. As all neighborhoods are independent of each other, we only have to make sure that the relations of the different pieces corresponding to a single neighborhood do not violate the relations of the compound regions. This can be done independently for every neighborhood.

Consider a particular neighborhood $N(w)$. If w is not a boundary world, then only the relations PO, TPP, NTPP, and their converse are possible between the regions affiliated with $N(w)$, since they share $N(w)$ as their common interior. For closing the neighborhood $N(w)$, all pieces must fulfill the "part of" relations whereas the PO relations cannot be violated as long as the corresponding pieces have a common interior.

One possibility to fulfill the "part of" relations is using a hierarchy of the regions, defined as follows.

Definition 9.9 (hierarchy of regions).
A hierarchy of regions H_Θ is a mapping of regions to levels, where a region X is of level $H_\Theta(X) = 1$ if there is no region Y which is part of X. A region X is of level $H_\Theta(X) = k$ if there is a region Y of level $H_\Theta(Y) = k - 1$ which is part of X and if there is no region Z which is part of X and has a higher level than $H_\Theta(Z) = k - 1$.[7]

The pieces of all regions affiliated with $N(w)$ must then be chosen according to H_Θ, i.e., pieces of regions of the same level are equal for this particular neighborhood and are non-tangential proper part of all pieces of regions of a higher level. We choose the single pieces to be rectangles.

If w is a boundary world, the boundary regions of w are only affiliated with one part of $N(w)$ and their pieces must share the common boundary. Therefore, both parts of $N(w)$ must be closed separately according to H_Θ (see Figure 9.4c). In the same way as for two-dimensional space, neighborhoods can be placed in any higher dimensional space and closed therein according to H_Θ. As the three points corresponding to a world w of level 0 and its two R_I-successors can always be aligned, $N(w)$ can also be placed on a line. Thus, all neighborhoods can be placed independently in a one-dimensional space and closed as intervals according to H_Θ (see Figure 9.3b).

Theorem 9.10. *Every consistent set of RCC-8 constraints can be realized in any dimension $d \geq 1$ where regions are (sets of) d-dimensional polytopes.*

---

[7] This corresponds to the finish time of depth-first search for each vertex of a graph G_Θ where regions are vertices V_Θ and "part of" relations are directed edges E_Θ, computable in time $O(V_\Theta + E_\Theta)$ [32, p.477ff]

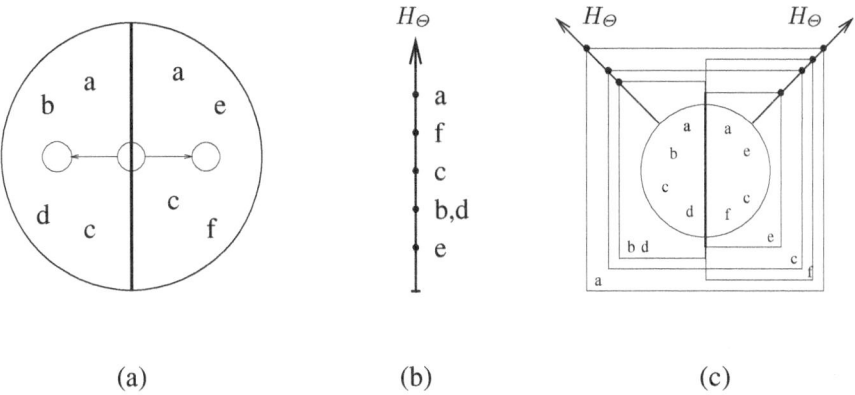

(a)   (b)   (c)

Fig. 9.4. (a) shows the two-dimensional neighborhood of a boundary world which is divided into two parts by the common boundary of the boundary regions b, d, e, and f. (b) shows a possible hierarchy H_Θ of regions. In (c) the neighborhood is closed with respect to H_Θ

So far all regions consist of as many pieces as there are neighborhoods affiliated with them, i.e., $O(n^2)$ many pieces for every region. We can further show that for three- and higher-dimensional space all regions can also be realized as internally connected. For this we construct a $d+1$-dimensional realization of internally connected regions by connecting all pieces of the same regions of a d-dimensional realization of internally disconnected regions.

Theorem 9.11. *Every consistent set* Θ *of* RCC-8 *constraints can be realized with internally connected regions in any dimension* $d \geq 3$ *where regions are polytopes.*

Proof. Suppose that Θ is consistent. With the following construction we obtain a three-dimensional realization of internally connected regions starting from a two-dimensional realization. We begin with constructing a particular two-dimensional realization in the plane determined by the x- and y-axes. (1a) Place all neighborhoods on a circle with center C such that the common boundary of each neighborhood corresponding to a boundary world points to C (see Figure 9.5a). (1b) Close all neighborhoods according to the hierarchy H_Θ such that all pieces of regions are rectangles. We now extend this two-dimensional realization to a three-dimensional realization. The third dimension is determined by the z-axis. (2a) Proceed from the two-dimensional realization according to H_Θ by first choosing pairwise distinct intervals on the positive z-axis for every region with $H_\Theta = 1$, i.e., for the regions that do not contain any other region. (2b) Build a pipe parallel to the z-axis for every piece of these regions starting from the plane ($z=0$) up to the endpoint of the corresponding interval. (2c) Connect the pipes of the same region within the range of the corresponding interval using pipes pointing to the center (see

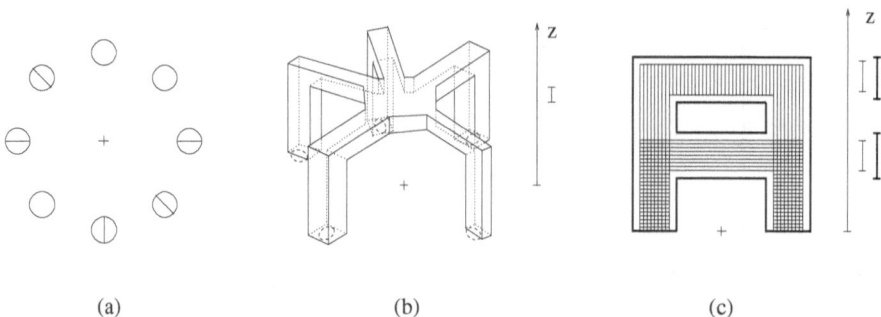

Fig. 9.5. Construction of the three-dimensional realization. (a) placing the two-dimensional neighborhoods on a circle. (b) connecting the pieces of a region on a particular level. (c) connecting the pipes of a region (bold line) that contains the vertically and the horizontally hashed regions

Figure 9.5b). (3) Next the regions with $H_\Theta = 2$ are connected, i.e., those regions that only contain already connected regions. To do this, (3a) choose intervals on the z-axis for these regions such that the intervals contain all intervals of the contained regions but do not overlap with any other interval. (3b) Build a pipe for every piece up to the endpoint of the corresponding interval with the largest z-value, and (3c) connect the pipes of every region within the range of all corresponding intervals (see Figure 9.5c). (4) Repeat step 3 successively for every level of H_Θ until all regions are connected. (5) Finally, close all neighborhoods on the negative z-axis according to H_Θ.

Obviously, with this construction all regions are internally connected. Furthermore all internally connected three-dimensional regions hold the same base relations as the two-dimensional realizations from which we started the construction. This is because all intervals on the z-axis are either contained in each other or are distinct, they have no common boundary points. All intervals corresponding to region X are contained in the intervals of region Y if and only if NTPP(x, y) or TPP(x, y). When two regions are disconnected they remain disconnected as they are not affiliated with the same neighborhoods. Two externally connected regions remain externally connected because every neighborhood was placed on the circle such that the common boundary points to its center. Therefore, if two of these regions are both affiliated with the same neighborhood, their pipes are externally connected and the horizontal connection of the single pipes is distinct. All other requirements of relations as, e.g., a common boundary point are already met by the pipes.

With a similar construction, a $d + 1$-dimensional realization of internally connected regions can be obtained from a d-dimensional realization of internally disconnected regions. In this case, the d-dimensional neighborhoods must be placed on a d-dimensional sphere and the intervals must be chosen at the $d + 1$ dimensional axis. All constructions kept the polytopic shape of the regions, so every region can be realized as a (d-dimensional) polytope. ∎

The restriction of regions to be polytopes can immediately be generalized to an arbitrary shape of regions.

9.4 Applicability of the Canonical Model

In the previous sections we reported about the existence of (reduced) RCC-8-models and how they can be mapped to topological spaces of different dimensions. In this section we study how RCC-8-models can be determined and how a realization can be generated from them. As there is a (reduced) RCC-8-model $\mathcal{M}$ for every consistent set of spatial relations Θ, and as it is always possible to generate a realization of $\mathcal{M}$, RCC-8 models are suitable for representing spatial regions with respect to their relations. RCC-8-models represent the characteristic points and information about their neighborhoods of a possible realization. As in the previous section we assume that Θ contains only constraints over the base relations.

9.4.1 Determination of RCC-8 Models

Given a set of RCC-8 constraints Θ which contains only base relations, we have to find a reduced RCC-8-model $\mathcal{M}$ for the modal encoding of RCC-8 such that only those worlds are boundary worlds of regions which are forced to be by the constraints. The Kripke frame of $\mathcal{M}$, i.e., the number of worlds and their accessibility relations are already known from the entailment constraints, but we have to find a valuation for every world and every region. For some worlds and some regions the valuation is already given by the constraints, for some it can be inferred using the constraints, for others it can be chosen. In order to make the inference step as easy as possible, we use the propositional encoding of RCC-8 with respect to a Kripke frame where every world w and every spatial variable x is transformed to a propositional atom X_w which is true if and only if X holds in w (see Section 6.2.3). The valuation of $\mathcal{M}$ can then be obtained from the satisfying assignment of the propositional formula. Although the encoding of the reduced RCC-8-models is not a Horn formula,[8] unit-resolution plus additional choices is sufficient for finding a satisfying assignment. As all clauses of the propositional encoding use worlds of the same RCC-8-cluster, the inference step is independent for every cluster. From Proposition 9.5 it is known which RCC-8-clusters contain a boundary world. Suppose that an RCC-8-cluster contains a boundary world, then the valuation of the two regions which introduced the boundary world can be

---

[8] This is because of the constraint $\Box(X \rightarrow \neg I \neg IX)$ which is transformed to $\bigwedge_{w \in W_0} (\neg X_w \lor X_w^1 \lor X_w^2)$.

chosen in all worlds of the RCC-8-cluster according to the relation of the two regions. The valuations of the other regions are either determined by unit-resolution or can be chosen according to their other valuations: If the valuation of a particular region in some world of the RCC-8-cluster is true, then the other valuations are also chosen as true, otherwise all valuations are chosen as false. If an RCC-8-cluster does not contain a boundary world, all worlds of the RCC-8-cluster have the same valuation. If the valuation of a region is not determined by unit-resolution it is chosen as false. With these choices a satisfying assignment is always found, even though the propositional formula is not Horn. As there are $O(n^2)$ worlds and n regions, there are $O(n^4)$ clauses (see Section 6.2.3), so a reduced RCC-8-model can be determined in time $O(n^4)$.

9.4.2 Generating a Realization

Suppose we have given a reduced RCC-8-model of a consistent set of RCC-8 constraints Θ. We have to distinguish the tasks of generating a realization of internally connected and disconnected regions. A realization of disconnected regions in d-dimensional space can be obtained by placing the $O(n^2)$ different neighborhoods in the d-dimensional space and close each neighborhood as specified in Section 9.3. For this, the hierarchy H_Θ of regions must be known, which can be computed in time $O(n + P_\Theta)$ where $P_\Theta \in O(n^2)$ is the number of "part of" relations in Θ (see Definition 9.9). Let $A_\Theta \in O(n)$ be the maximal number of regions affiliated with a neighborhood, then the closure of a neighborhood can be computed in time $O(A_\Theta)$.

Theorem 9.12. *Given a reduced* RCC-8 *model of a set of* RCC-8 *relations Θ, a realization of Θ in d-dimensional space ($d \geq 1$) can be generated in time $O(n^2 A_\Theta)$ when regions are allowed to be disconnected.*

In order to generate a realization of internally connected regions we can use the construction of the proof to Theorem 9.11. For every region we have to find the corresponding intervals on the z-axis. The number of intervals of a particular region X is equal to the number of regions with $H_\Theta = 1$ that are contained in X. Let $I_\Theta \in O(n)$ be the maximal number of regions with $H_\Theta = 1$ that are contained in a region.

Theorem 9.13. *Given a reduced* RCC-8 *model of a set of* RCC-8 *relations Θ, a realization of Θ with internally connected regions in d-dimensional space ($d \geq 3$) can be generated in time $O(n^2 A_\Theta I_\Theta)$.*

If $P_\Theta \in O(n^2)$ is the maximal number of neighborhoods affiliated with a region, every region can be realized as a polytope with $O(P_\Theta I_\Theta)$ vertices.

9.5 Discussion

In this chapter we identified a canonical model of RCC-8 based on Kripke semantics. In order to obtain a "topological" canonical model, we gave a topological interpretation of the Kripke models such that regions can be represented by points in the topological space and information about the neighborhood of these points with respect to the spatial relations holding between the regions. Using this canonical model, we proved that every consistent set of RCC-8 constraints has a realization in any dimension when regions are not forced to be internally connected, which is the case for regions as used by RCC-8. If regions are forced to be internally connected, we proved that a realization can always be found for three- and higher dimensional space. Furthermore, we give for the first time algorithms for generating realizations of either internally connected or disconnected regions.

There is some work on identifying canonical models for the RCC axioms, i.e., determining what mathematical structures fulfill all the RCC axioms, as, e.g., every region has a non-tangential proper part [138]. Gotts [72] found that every connected and regular topological space is a model for the RCC axioms. Stell and Worboys [152] identified a whole class of models based on Heyting structures. Both approaches only describe models for the RCC axioms, i.e., what kind of regions can be used at all. When additional constraints expressing relationships between regions are added, these results do not say anything about models anymore.[9] They are also by no means constructive, as they do not provide a way of effectively representing regions or generating realizations.

Previous approaches to dealing with dimension and internal connectedness of regions tried to specify predicates and suitable axioms in order to restrict dimension and connectedness of regions [11, 71]. As all regular regions have the same dimension as the underlying space, using our results it is not necessary for consistency purposes to specify the dimension of regions explicitly if internally disconnected regions are permitted. If internally connected regions are required, these predicates only have an influence on the consistency of a set of spatial relations in one- or two-dimensional applications. In three- and higher-dimensional space all regions may be either internally connected or disconnected. Forcing internal connectedness of regions in two-dimensional space leads to difficult computational problems as there are no algorithms for dealing with this task. As Grigni et al. [77] pointed out, a well-known open problem in graph theory which is NP-hard but not known to be decidable [105, 106] can be reduced to the consistency problem for two-dimensional internally connected regions.

It is certainly the better approach to have an additional connectedness predicate than forcing all regions to be internally connected which is done,

---

[9] Consider the hypothetical case of a set of RCC-8 constraints which is realizable in two- but not in one-dimensional space. Then, a one-dimensional space is still a model of the RCC axioms, but not of RCC-8.

e.g., by the similar calculus of Egenhofer [44], as there are many applications where regions are in fact disconnected. Within the area of geographical information systems, e.g., which offer a great variety of possible applications, many countries or other geographical entities are not internally connected regions. In areas like computer vision one often deals with two-dimensional projections of the three-dimensional space where many connected objects are perceived as disconnected objects due to occlusion. In robot navigation, maps are often two-dimensional cuttings of a three-dimensional space.

10. Conclusions

In this work we have provided a comprehensive analysis of the Region Connection Calculus RCC-8, a well-known topological constraint language for qualitative spatial representation and reasoning. Our work was motivated by the general research goal of qualitative spatial representation and reasoning, namely, by developing an expressive spatial calculus which covers many different aspects of space, which allows for efficient reasoning, and which can be regarded as cognitively adequate. Since topological relationships are a very fundamental aspect of space and since RCC-8 is provided with formal semantics which allow for a theoretical analysis of the calculus, RCC-8 seemed to be an excellent starting point for reaching this goal.

10.1 Summary of Contributions

The contributions of this work are threefold. They are concerned with cognitive, with computational, as well as with representational aspects of RCC-8. In all three areas, we investigated interesting open problems which resulted from previous research on RCC-8 in particular and on qualitative spatial reasoning in general and whose solutions help to approach the general research goal of qualitative spatial reasoning.

Cognitive Properties of RCC8

Cognitive adequacy is an important motivation for qualitative spatial reasoning. However, most often this relies on the researchers' intuition rather than on empirical data. In a series of empirical investigations we tried to evaluate the cognitive adequacy of the topological distinctions made by RCC-8. We found that subjects distinguish the spatial configurations given in our grouping tasks mainly by their topological relationships (in particular by those distinctions made by RCC-8) and that distinctions according to other spatial aspects such as size, distance, or direction are only made as a further refinement of the topological relations. These results give strong empirical evidence of the (conceptual) cognitive adequacy of RCC-8 and of the importance of topological relationships in human spatial cognition.

J. Renz: Qualitative Spatial Reasoning with Topological Information, LNAI 2293, pp. 173–177, 2002.
© Springer-Verlag Berlin Heidelberg 2002

Computational Properties of RCC8

Efficient reasoning mechanisms are fundamental for many applications of qualitative spatial reasoning. Since NP-hard reasoning problems do not allow for efficient reasoning methods in general, a detailed analysis of the computational properties of the problem is in many cases indispensable for developing algorithms which solve most problem instances efficiently. While the computational properties of qualitative temporal reasoning have been studied for a long time, there are almost no results on the computational properties of qualitative spatial reasoning. This is mainly due to the multi-dimensional nature of space which makes it much more difficult to give formal semantics of a calculus and to study its formal properties. As a first step of our computational analysis of RCC-8, we proved NP-hardness of the consistency problem of RCC-8, which is the fundamental reasoning problem of relation-based reasoning. This leads to the question of whether there are subsets of RCC-8 for which the consistency problem is polynomial. The goal of such an analysis is to identify maximal tractable subsets which contain all base relations. These sets can be used to speed up reasoning over the general NP-hard consistency problem by reducing the search space of the backtracking search. Using a modification of Bennett's encoding of RCC-8 in modal logic, we transformed reasoning over RCC-8 to computing propositional satisfiability and identified those RCC-8 relations that are transformed to propositional Horn formulas. The consistency problem of this set of 148 RCC-8 relations (which we called $\widehat{\mathcal{H}}_8$) is tractable, and by a computer assisted case-analysis and additional NP-hardness proofs we showed that $\widehat{\mathcal{H}}_8$ is also a maximal tractable subset of RCC-8. We further showed that enforcing path-consistency is complete for deciding consistency of constraints over $\widehat{\mathcal{H}}_8$.

Having identified the maximal tractable subset $\widehat{\mathcal{H}}_8$, it is not clear whether $\widehat{\mathcal{H}}_8$ is the only maximal tractable subset that contains all base relations of RCC-8 or whether there are other such sets. We solved this problem by developing a general method for deciding tractability of reasoning over systems of relations. This method takes as input a "refinement strategy" of relations of a certain set $\mathcal{S}$ to relations of a set $\mathcal{T}$ for which path-consistency is known to decide consistency, and checks whether every path-consistent set of constraints over $\mathcal{S}$ can be consistently refined to a set of constraints over $\mathcal{T}$ when using the refinement strategy. If the refinement strategy is valid, path-consistency decides consistency of sets of constraints over $\mathcal{S}$, and, hence, reasoning with $\mathcal{S}$ is tractable. We proposed a simple heuristic for finding a suitable refinement strategy which was successful for all tractable subsets of RCC-8 and of Allen's interval algebra that contain all base relations. Using this method and a computer assisted generation of all possible candidates, we made a complete analysis of tractability of RCC-8 by identifying all maximal tractable subsets of RCC-8 that contain all base relations—altogether three maximal tractable subsets. We also gave the fastest possible algorithms for finding a consistent scenario of any path-consistent set of constraints over each one of the three maximal tractable subsets.

The maximal tractable subsets give the possibility to solve the general NP-hard consistency problem of RCC-8 more efficiently than previously possible. In an empirical evaluation with randomly generated instances using all RCC-8 relations, we found that almost all instances of the NP-hard problem we generated—even those in the phase-transition region—up to a size of $n = 100$ regions can be solved very efficiently when using the method of orthogonally combining different heuristics. In particular those heuristics involving the maximal tractable subsets are very successful. They solved more instances than the other heuristics in reasonable time and also solved the instances faster in general. Using an orthogonal combination of the most successful heuristics (which all involve the maximal tractable subsets), we found that almost all of our randomly generated instances can be solved within a given maximal number of visited nodes in the search space which is linear in the size n of the instances. Up to a size of about $n = 400$ regions this resulted in an acceptable running time. When randomly generating instances using only relations which are not contained in any of the maximal tractable subsets, the instances are in general much harder to solve than instances using all RCC-8 relations. In particular instances of the phase-transition region can become very hard, many of them cannot be solved within reasonable time when they are larger than $n = 60$ regions. However, the heuristics using the maximal tractable subsets still solve more instances and in less time than the other heuristics.

Representational Properties of RCC8

Spatial reasoning is mainly applied in two- or three-dimensional Euclidean space rather than in arbitrary dimensional topological space as used by RCC-8. An important problem is therefore whether RCC-8 can be used for Euclidean space of a particular dimension, e.g., whether a consistent set of constraints over RCC-8 can be realized in this kind of space. Related to this question is the problem of representing arbitrary spatial regions as used by RCC-8 within a computational framework. We solved these problems by determining a canonical model of RCC-8 by which every consistent set of constraints over RCC-8 can be represented. Using a transformation of the canonical model to topological space, we were able to prove that any consistent set can be realized in topological space of any dimension (in particular also Euclidean space) if regions are allowed to consist of multiple pieces which is the case for RCC-8. If regions are required to be internally connected, i.e., one-piece regions, we showed that every consistent set can be represented in topological space of dimension $d \geq 3$.

10.2 Discussion & Future Research

These results establish RCC-8 as an excellent basis for reaching the general research goal of qualitative spatial reasoning. Our results on cognitive adequacy of RCC-8 do not only show that the distinctions made by RCC-8 are an important part of human spatial knowledge and, hence, justify the importance of RCC-8 for qualitative spatial reasoning. They also show that it is reasonable to choose RCC-8 as a starting point and extend RCC-8 by other spatial aspects. This is further confirmed by our formal results. Our canonical model of RCC-8 and its transformations show that RCC-8 can be used for topological spaces of any dimension; in particular also for two- and three-dimensional Euclidean space, the most commonly used notion of space. This makes RCC-8 very general and widely applicable. As shown in a detailed empirical analysis, our complete analysis of tractability of RCC-8 enables surprisingly efficient solutions to the NP-complete reasoning problems of RCC-8.

Ergo, all conditions of using RCC-8 as a starting point for reaching the general goal of qualitative spatial reasoning have been fulfilled. The next step towards approaching this goal is to extend RCC-8 by other spatial aspects such as direction, distance, or size. These extensions should all use a common ontology of space, e.g., they should all use regions as the basic entities instead of points. For some spatial aspects it will therefore be necessary to develop new calculi. This should be done in a way that the calculi for treating different spatial aspects itself fulfill all necessary conditions, i.e., they should have formal semantics which enables a computational analysis and which allows to combine them with other calculi. Also, depending on the desired applications, they should preferably be cognitively adequate. For the task of finding large tractable subsets of the calculi in order to obtain more efficient reasoning mechanisms, the method we developed in Chapter 7 should be of great help. Apart from these requirements of the single calculi, it will be important to analyze the possible interactions and interdependencies between the different calculi, and to develop algorithms for handling them.

As a first extension of RCC-8 by other spatial aspects, we have added qualitative information about the size of regions [68]. The relative size constraints we have used are of the form $size(x)Rsize(y)$ where $size(.)$ specifies the size of a region and R is a disjunction over the base relations $<, >$, and $=$. These relations correspond to the point algebra [160] and have well studied properties. Topological and qualitative size constraints are not independent of each other. Given for instance the RCC-8 constraint $x\{\mathsf{TPP}\}y$, then $size(x)$ must be smaller than $size(y)$. Conversely, the size constraint $size(x)\{<, =\}size(y)$, for instance, enforces the RCC-8 constraint $x\{\mathsf{DC}, \mathsf{EC}, \mathsf{PO}, \mathsf{TPP}, \mathsf{NTPP}, \mathsf{EQ}\}y$.

We developed an $O(n^3)$ time algorithm BIPATH-CONSISTENCY which takes as input a set of RCC-8 constraints Θ and a set of qualitative size constraints Σ and which enforces path-consistency to both sets while propagating information from one set to the other and *vice versa*. We proved that when Θ

contains only relations of the maximal tractable subset $\widehat{\mathcal{H}}_8$ and after applying BIPATH-CONSISTENCY to Θ and Σ both sets are path-consistent, then both sets together are consistent. This property can easily be transfered to the other two maximal tractable subsets $\mathcal{C}_8$ and $\mathcal{Q}_8$. Furthermore, if Σ contains only constraints over the base relations, then BIPATH-CONSISTENCY is sufficient for deciding consistency of the combined sets even if the topological set alone is an instance of an NP-complete problem. This is because the propagation of definite size constraints to topological constraints refines all RCC-8 constraints to constraints over $\widehat{\mathcal{H}}_8$. Thus, the addition of qualitative size information to topological information as given by RCC-8 does not negatively effect the computational properties of RCC-8.

Certainly, it cannot be expected that extensions of RCC-8 by other spatial aspects such as direction or distance are as well-behaved as the extension by relative size information, but it will definitely be an interesting and rewarding challenge to tackle these extensions.

A. Enumeration of the Relations of the Maximal Tractable Subsets of RCC-8

In this appendix we enumerate all relations of RCC-8, their membership in certain subsets of RCC-8 and some properties of the relations. In Section A.1 we list all relations of the set $\mathcal{H}_8$ together with their abbreviated form as given in Definition 6.12. In Section A.2 we list all relations which are contained in $\widehat{\mathcal{H}}_8$ but not in $\mathcal{H}_8$ and give their functional construction using relations of $\mathcal{H}_8$ according to Lemma 6.50. In Section A.3 we list all relations of RCC-8 and give their membership in the maximal tractable subsets.

A.1 Relations of $\mathcal{H}_8$ and Their Abbreviated Form

Relations of $\mathcal{H}_8$	Abbreviated Form
$\{DC\}$	δ
$\{EC\}$	$\neg\delta \wedge \delta i$
$\{DC, EC\}$	δi
$\{PO\}$	$\neg\pi \wedge \neg\gamma \wedge \neg\delta i$
$\{DC, PO\}$	$\neg\pi \wedge \neg\gamma \wedge (\delta \vee \neg\delta i)$
$\{EC, PO\}$	$\neg\delta \wedge \neg\pi \wedge \neg\gamma$
$\{DC, EC, PO\}$	$\neg\pi \wedge \neg\gamma$
$\{TPP\}$	$\pi \wedge \neg\gamma \wedge \neg\pi i$
$\{DC, TPP\}$	$\neg\gamma \wedge \neg\pi i \wedge (\delta \vee \pi)$
$\{EC, TPP\}$	$\neg\delta \wedge \neg\gamma \wedge \neg\pi i \wedge (\delta i \vee \pi)$
$\{DC, EC, TPP\}$	$\neg\gamma \wedge \neg\pi i \wedge (\delta i \vee \pi)$
$\{PO, TPP\}$	$\neg\gamma \wedge \neg\delta i \wedge \neg\pi i$
$\{DC, PO, TPP\}$	$\neg\gamma \wedge \neg\pi i \wedge (\delta \vee \neg\delta i)$
$\{EC, PO, TPP\}$	$\neg\delta \wedge \neg\gamma \wedge \neg\pi i$
$\{DC, EC, PO, TPP\}$	$\neg\gamma \wedge \neg\pi i$
$\{NTPP\}$	$\neg\gamma \wedge \pi i$
$\{DC, NTPP\}$	$\neg\gamma \wedge (\delta \vee \pi i)$
$\{EC, NTPP\}$	$\neg\delta \wedge \neg\gamma \wedge (\delta i \vee \pi i)$
$\{DC, EC, NTPP\}$	$\neg\gamma \wedge (\delta i \vee \pi i)$
$\{TPP, NTPP\}$	$\pi \wedge \neg\gamma$

J. Renz: Qualitative Spatial Reasoning with Topological Information, LNAI 2293, pp. 179–190, 2002.
© Springer-Verlag Berlin Heidelberg 2002

Relations of $\mathcal{H}_8$	Abbreviated Form
$\{DC, TPP, NTPP\}$	$\neg\gamma \wedge (\delta \vee \pi)$
$\{EC, TPP, NTPP\}$	$\neg\delta \wedge \neg\gamma \wedge (\delta i \vee \pi)$
$\{DC, EC, TPP, NTPP\}$	$\neg\gamma \wedge (\delta i \vee \pi)$
$\{PO, TPP, NTPP\}$	$\neg\gamma \wedge \neg\delta i$
$\{DC, PO, TPP, NTPP\}$	$\neg\gamma \wedge (\delta \vee \neg\delta i)$
$\{EC, PO, TPP, NTPP\}$	$\neg\delta \wedge \neg\gamma$
$\{DC, EC, PO, TPP, NTPP\}$	$\neg\gamma$
$\{TPP^{-1}\}$	$\neg\pi \wedge \gamma \wedge \neg\gamma i$
$\{DC, TPP^{-1}\}$	$\neg\pi \wedge \neg\gamma i \wedge (\delta \vee \gamma)$
$\{EC, TPP^{-1}\}$	$\neg\delta \wedge \neg\pi \wedge \neg\gamma i \wedge (\delta i \vee \gamma)$
$\{DC, EC, TPP^{-1}\}$	$\neg\pi \wedge \neg\gamma i \wedge (\delta i \vee \gamma)$
$\{PO, TPP^{-1}\}$	$\neg\pi \wedge \neg\delta i \wedge \neg\gamma i$
$\{DC, PO, TPP^{-1}\}$	$\neg\pi \wedge \neg\gamma i \wedge (\delta \vee \neg\delta i)$
$\{EC, PO, TPP^{-1}\}$	$\neg\delta \wedge \neg\pi \wedge \neg\gamma i$
$\{DC, EC, PO, TPP^{-1}\}$	$\neg\pi \wedge \neg\gamma i$
$\{NTPP^{-1}\}$	$\neg\pi \wedge \gamma i$
$\{DC, NTPP^{-1}\}$	$\neg\pi \wedge (\delta \vee \gamma i)$
$\{EC, NTPP^{-1}\}$	$\neg\delta \wedge \neg\pi \wedge (\delta i \vee \gamma i)$
$\{DC, EC, NTPP^{-1}\}$	$\neg\pi \wedge (\delta i \vee \gamma i)$
$\{TPP^{-1}, NTPP^{-1}\}$	$\neg\pi \wedge \gamma$
$\{DC, TPP^{-1}, NTPP^{-1}\}$	$\neg\pi \wedge (\delta \vee \gamma)$
$\{EC, TPP^{-1}, NTPP^{-1}\}$	$\neg\delta \wedge \neg\pi \wedge (\delta i \vee \gamma)$
$\{DC, EC, TPP^{-1}, NTPP^{-1}\}$	$\neg\pi \wedge (\delta i \vee \gamma)$
$\{PO, TPP^{-1}, NTPP^{-1}\}$	$\neg\pi \wedge \neg\delta i$
$\{DC, PO, TPP^{-1}, NTPP^{-1}\}$	$\neg\pi \wedge (\delta \vee \neg\delta i)$
$\{EC, PO, TPP^{-1}, NTPP^{-1}\}$	$\neg\delta \wedge \neg\pi$
$\{DC, EC, PO, TPP^{-1}, NTPP^{-1}\}$	$\neg\pi$
$\{EQ\}$	$\pi \wedge \gamma$
$\{DC, EQ\}$	$(\delta \vee \pi) \wedge (\delta \vee \gamma)$
$\{EC, EQ\}$	$\neg\delta \wedge (\delta i \vee \pi) \wedge (\delta i \vee \gamma)$
$\{DC, EC, EQ\}$	$(\delta i \vee \pi) \wedge (\delta i \vee \gamma)$
$\{TPP, NTPP, EQ\}$	π
$\{DC, TPP, NTPP, EQ\}$	$(\delta \vee \pi)$
$\{EC, TPP, NTPP, EQ\}$	$\neg\delta \wedge (\delta i \vee \pi)$
$\{DC, EC, TPP, NTPP, EQ\}$	$(\delta i \vee \pi)$
$\{TPP^{-1}, NTPP^{-1}, EQ\}$	γ
$\{DC, TPP^{-1}, NTPP^{-1}, EQ\}$	$(\delta \vee \gamma)$
$\{EC, TPP^{-1}, NTPP^{-1}, EQ\}$	$\neg\delta \wedge (\delta i \vee \gamma)$
$\{DC, EC, TPP^{-1}, NTPP^{-1}, EQ\}$	$(\delta i \vee \gamma)$
$\{PO, TPP, NTPP, TPP^{-1}, NTPP^{-1}, EQ\}$	$\neg\delta i$
$\{DC, PO, TPP, NTPP, TPP^{-1}, NTPP^{-1}, EQ\}$	$(\delta \vee \neg\delta i)$
$\{EC, PO, TPP, NTPP, TPP^{-1}, NTPP^{-1}, EQ\}$	$\neg\delta$

A.2 Functional Construction of the Relations of $\widehat{\mathcal{H}}_8 \setminus \mathcal{H}_8$ Using Relations of $\mathcal{H}_8$

Here we give the functional construction of the relations of $\widehat{\mathcal{H}}_8 \setminus \mathcal{H}_8$ using relations of $\mathcal{H}_8$ according to Lemma 6.50. In order to save space, we abbreviate some relations by writing them as the complement $\overline{R}$ of other relations R (i.e., $\overline{R}$ contains all base relations not included in R).

Relations of $\widehat{\mathcal{H}}_8 \setminus \mathcal{H}_8$	Functional Construction
$\overline{\{EQ\}}$	$\{EC\} \circ \{DC, PO\}$
$\{EC, PO, TPP, NTPP, EQ\}$	$\{EC, TPP, NTPP, EQ\} \circ \{TPP, NTPP, EQ\}$
$\overline{\{TPP^{-1}, NTPP^{-1}\}}$	$\{DC, TPP, NTPP, EQ\} \circ \{TPP, NTPP, EQ\}$
$\overline{\{NTPP, NTPP^{-1}\}}$	$\{EC\} \circ \{EC\}$
$\overline{\{DC, EC, NTPP^{-1}\}}$	$\{PO, TPP^{-1}\} \circ \{TPP\}$
$\overline{\{DC, NTPP^{-1}\}}$	$\{EC, PO, TPP^{-1}\} \circ \{TPP\}$
$\overline{\{NTPP^{-1}\}}$	$\{EC\} \circ \{EC, PO\}$
$\overline{\{NTPP\}}$	$\{EC\} \circ \{DC, EC\}$
$\{DC, EC, PO, TPP, TPP^{-1}\}$	$\overline{\{EQ\}} \cap \overline{\{NTPP, NTPP^{-1}\}}$
$\{PO, TPP, NTPP, TPP^{-1}\}$	$\overline{\{EQ\}} \cap \overline{\{DC, EC, NTPP^{-1}\}}$
$\overline{\{DC, NTPP^{-1}, EQ\}}$	$\overline{\{EQ\}} \cap \overline{\{DC, NTPP^{-1}\}}$
$\overline{\{NTPP^{-1}, EQ\}}$	$\overline{\{EQ\}} \cap \overline{\{NTPP^{-1}\}}$
$\overline{\{NTPP, EQ\}}$	$\overline{\{EQ\}} \cap \overline{\{NTPP\}}$
$\overline{\{DC, EC, EQ\}}$	$\overline{\{EQ\}} \cap \overline{\{DC, EC\}}$
$\overline{\{EC, EQ\}}$	$\overline{\{EQ\}} \cap \overline{\{EC\}}$
$\overline{\{DC, EQ\}}$	$\overline{\{EQ\}} \cap \overline{\{DC\}}$
$\{TPP, EQ\}$	$\{TPP, NTPP, EQ\} \cap \overline{\{NTPP, NTPP^{-1}\}}$
$\{DC, TPP, EQ\}$	$\{DC, TPP, NTPP, EQ\} \cap \overline{\{NTPP, NTPP^{-1}\}}$
$\{EC, TPP, EQ\}$	$\{EC, TPP, NTPP, EQ\} \cap \overline{\{NTPP, NTPP^{-1}\}}$
$\{DC, EC, TPP, EQ\}$	$\{DC, EC, TPP, NTPP, EQ\} \cap \overline{\{NTPP, NTPP^{-1}\}}$
$\{EC, PO, TPP, EQ\}$	$\{EC, PO, TPP, NTPP, EQ\} \cap \overline{\{NTPP, NTPP^{-1}\}}$
$\{DC, EC, PO, TPP, EQ\}$	$\overline{\{TPP^{-1}, NTPP^{-1}\}} \cap \overline{\{NTPP, NTPP^{-1}\}}$
$\{PO, TPP, NTPP, EQ\}$	$\{EC, PO, TPP, NTPP, EQ\} \cap \overline{\{DC, EC, NTPP^{-1}\}}$
$\{DC, PO, TPP, NTPP, EQ\}$	$\overline{\{TPP^{-1}, NTPP^{-1}\}} \cap \overline{\{EC\}}$
$\{PO, TPP, TPP^{-1}, EQ\}$	$\overline{\{NTPP, NTPP^{-1}\}} \cap \overline{\{DC, EC, NTPP^{-1}\}}$
$\{DC, PO, TPP, TPP^{-1}, EQ\}$	$\overline{\{NTPP, NTPP^{-1}\}} \cap \overline{\{EC\}}$

Relations of $\widehat{\mathcal{H}}_8 \setminus \mathcal{H}_8$	Functional Construction
$\{EC, PO, TPP, TPP^{-1}, EQ\}$	$\{NTPP, NTPP^{-1}\} \cap \{DC, NTPP^{-1}\}$
$\overline{\{EC, NTPP^{-1}\}}$	$\{NTPP^{-1}\} \cap \overline{\{EC\}}$
$\overline{\{DC, EC, NTPP\}}$	$\{NTPP\} \cap \{DC, EC\}$
$\overline{\{EC, NTPP\}}$	$\{NTPP\} \cap \{EC\}$
$\overline{\{DC, NTPP\}}$	$\{NTPP\} \cap \{DC\}$
$\{DC, PO, TPP, TPP^{-1}\}$	$\{DC, EC, PO, TPP, TPP^{-1}\} \cap \overline{\{EC, EQ\}}$
$\{EC, PO, TPP, TPP^{-1}\}$	$\{DC, EC, PO, TPP, TPP^{-1}\} \cap \overline{\{DC, NTPP^{-1}, EQ\}}$
$\{PO, TPP, TPP^{-1}\}$	$\{DC, PO, TPP, TPP^{-1}\} \cap \{EC, PO, TPP, TPP^{-1}\}$
$\overline{\{EC, NTPP^{-1}, EQ\}}$	$\{NTPP^{-1}, EQ\} \cap \overline{\{EC, EQ\}}$
$\overline{\{DC, EC, NTPP, EQ\}}$	$\overline{\{NTPP, EQ\}} \cap \overline{\{DC, EC, EQ\}}$
$\overline{\{EC, NTPP, EQ\}}$	$\overline{\{NTPP, EQ\}} \cap \overline{\{EC, EQ\}}$
$\overline{\{DC, NTPP, EQ\}}$	$\overline{\{NTPP, EQ\}} \cap \overline{\{DC, EQ\}}$
$\{PO, TPP, EQ\}$	$\{EC, PO, TPP, EQ\} \cap \{PO, TPP, NTPP, EQ\}$
$\{DC, PO, TPP, EQ\}$	$\{DC, EC, PO, TPP, EQ\}$ $\cap \{DC, PO, TPP, NTPP, EQ\}$
$\overline{\{TPP, EQ\}}$	$\{EC, NTPP\} \circ \{DC, EQ\}$
$\overline{\{TPP^{-1}, EQ\}}$	$\{DC, EQ\} \circ \{EC, NTPP^{-1}\}$
$\overline{\{DC, TPP, NTPP\}}$	$\{TPP^{-1}, NTPP^{-1}, EQ\} \circ \{EC, EQ\}$
$\overline{\{TPP, NTPP\}}$	$\{EC, EQ\} \circ \{DC, EQ\}$
$\{PO, NTPP\}$	$\{PO, TPP, NTPP\} \cap \overline{\{TPP, EQ\}}$
$\{DC, PO, NTPP\}$	$\{DC, PO, TPP, NTPP\} \cap \overline{\{TPP, EQ\}}$
$\{EC, PO, NTPP\}$	$\{EC, PO, TPP, NTPP\} \cap \overline{\{TPP, EQ\}}$
$\{DC, EC, PO, NTPP\}$	$\{DC, EC, PO, TPP, NTPP\} \cap \overline{\{TPP, EQ\}}$
$\{PO, NTPP, TPP^{-1}\}$	$\{PO, TPP, NTPP, TPP^{-1}\} \cap \overline{\{TPP, EQ\}}$
$\{DC, PO, NTPP, TPP^{-1}\}$	$\overline{\{EC, NTPP^{-1}, EQ\}} \cap \overline{\{TPP, EQ\}}$
$\{EC, PO, NTPP, TPP^{-1}\}$	$\overline{\{DC, NTPP^{-1}, EQ\}} \cap \overline{\{TPP, EQ\}}$
$\{DC, EC, PO, NTPP, TPP^{-1}\}$	$\{NTPP^{-1}, EQ\} \cap \{TPP, EQ\}$
$\{PO, NTPP^{-1}\}$	$\{TPP^{-1}, EQ\} \cap \{PO, TPP^{-1}, NTPP^{-1}\}$
$\{DC, PO, NTPP^{-1}\}$	$\{TPP^{-1}, EQ\} \cap \{DC, PO, TPP^{-1}, NTPP^{-1}\}$
$\{EC, PO, NTPP^{-1}\}$	$\{TPP^{-1}, EQ\} \cap \{EC, PO, TPP^{-1}, NTPP^{-1}\}$
$\{DC, EC, PO, NTPP^{-1}\}$	$\{TPP^{-1}, EQ\} \cap \{DC, EC, PO, TPP^{-1}, NTPP^{-1}\}$
$\{PO, TPP, NTPP^{-1}\}$	$\{TPP^{-1}, EQ\} \cap \overline{\{DC, EC, NTPP, EQ\}}$
$\{DC, PO, TPP, NTPP^{-1}\}$	$\{TPP^{-1}, EQ\} \cap \overline{\{EC, NTPP, EQ\}}$
$\{EC, PO, TPP, NTPP^{-1}\}$	$\{TPP^{-1}, EQ\} \cap \overline{\{DC, NTPP, EQ\}}$
$\{DC, EC, PO, TPP, NTPP^{-1}\}$	$\{TPP^{-1}, EQ\} \cap \overline{\{NTPP, EQ\}}$

Relations of $\widehat{\mathcal{H}}_8 \setminus \mathcal{H}_8$	Functional Construction
$\overline{\{TPP, TPP^{-1}, EQ\}}$	$\{TPP^{-1}, EQ\} \cap \overline{\{TPP, EQ\}}$
$\{PO, TPP, NTPP, NTPP^{-1}\}$	$\{TPP^{-1}, EQ\} \cap \overline{\{DC, EC, EQ\}}$
$\overline{\{EC, TPP^{-1}, EQ\}}$	$\{TPP^{-1}, EQ\} \cap \overline{\{EC, EQ\}}$
$\overline{\{DC, TPP^{-1}, EQ\}}$	$\{TPP^{-1}, EQ\} \cap \overline{\{DC, EQ\}}$
$\overline{\{DC, EC, TPP, EQ\}}$	$\overline{\{TPP, EQ\}} \cap \{DC, EC, EQ\}$
$\overline{\{EC, TPP, EQ\}}$	$\overline{\{TPP, EQ\}} \cap \{EC, EQ\}$
$\overline{\{DC, TPP, EQ\}}$	$\overline{\{TPP, EQ\}} \cap \{DC, EQ\}$
$\{PO, EQ\}$	$\{PO, TPP, EQ\} \cap \overline{\{DC, NTPP, TPP\}}$
$\{DC, PO, EQ\}$	$\{DC, PO, TPP, EQ\} \cap \overline{\{TPP, NTPP\}}$
$\{EC, PO, EQ\}$	$\{EC, PO, TPP, EQ\} \cap \overline{\{DC, NTPP, TPP\}}$
$\{DC, EC, PO, EQ\}$	$\{DC, EC, PO, TPP, EQ\} \cap \overline{\{TPP, NTPP\}}$
$\{TPP^{-1}, EQ\}$	$\{PO, TPP, TPP^{-1}, EQ\} \cap \{TPP^{-1}, NTPP^{-1}, EQ\}$
$\{PO, TPP^{-1}, EQ\}$	$\{PO, TPP, TPP^{-1}, EQ\} \cap \overline{\{DC, NTPP, TPP\}}$
$\{DC, PO, TPP^{-1}, EQ\}$	$\{DC, PO, TPP, TPP^{-1}, EQ\} \cap \overline{\{TPP, NTPP\}}$
$\{EC, PO, TPP^{-1}, EQ\}$	$\{EC, PO, TPP, TPP^{-1}, EQ\} \cap \overline{\{DC, NTPP, TPP\}}$
$\{DC, EC, TPP^{-1}, EQ\}$	$\{NTPP, NTPP^{-1}\} \cap \overline{\{PO, TPP, NTPP\}}$
$\{DC, TPP^{-1}, EQ\}$	$\{DC, EC, TPP^{-1}, EQ\} \cap \{DC, TPP^{-1}, NTPP^{-1}, EQ\}$
$\{EC, TPP^{-1}, EQ\}$	$\{DC, EC, TPP^{-1}, EQ\} \cap \{EC, TPP^{-1}, NTPP^{-1}, EQ\}$
$\{DC, EC, PO, TPP^{-1}, EQ\}$	$\{NTPP, NTPP^{-1}\} \cap \overline{\{TPP, NTPP\}}$
$\{PO, TPP^{-1}, NTPP^{-1}, EQ\}$	$\overline{\{DC, NTPP, TPP\}} \cap \overline{\{DC, EC, NTPP\}}$
$\overline{\{EC, TPP, NTPP\}}$	$\overline{\{TPP, NTPP\}} \cap \overline{\{EC, NTPP\}}$
$\{PO, NTPP, NTPP^{-1}\}$	$\overline{\{TPP, TPP^{-1}, EQ\}} \cap \{PO, TPP, NTPP, NTPP^{-1}\}$
$\{DC, PO, NTPP, NTPP^{-1}\}$	$\overline{\{TPP, TPP^{-1}, EQ\}} \cap \overline{\{EC, TPP^{-1}, EQ\}}$
$\{EC, PO, NTPP, NTPP^{-1}\}$	$\overline{\{TPP, TPP^{-1}, EQ\}} \cap \overline{\{DC, TPP^{-1}, EQ\}}$

A.3 The Maximal Tractable Subsets of RCC-8

In this section we list all relations of RCC-8 and their membership in the three maximal tractable subsets $\widehat{\mathcal{H}}_8, \mathcal{C}_8,$ and $\mathcal{Q}_8$. A bullet ($\bullet$) indicates membership of the relation in the corresponding set.

Relation	$\widehat{\mathcal{H}}_8$	$\mathcal{C}_8$	$\mathcal{Q}_8$
{DC}	$\bullet$	$\bullet$	$\bullet$
{EC}	$\bullet$	$\bullet$	$\bullet$
{DC, EC}	$\bullet$	$\bullet$	$\bullet$
{PO}	$\bullet$	$\bullet$	$\bullet$
{DC, PO}	$\bullet$	$\bullet$	$\bullet$
{EC, PO}	$\bullet$	$\bullet$	$\bullet$
{DC, EC, PO}	$\bullet$	$\bullet$	$\bullet$
{TPP}	$\bullet$	$\bullet$	$\bullet$
{DC, TPP}	$\bullet$	$\bullet$	$\bullet$
{EC, TPP}	$\bullet$		$\bullet$
{DC, EC, TPP}	$\bullet$		$\bullet$
{PO, TPP}	$\bullet$	$\bullet$	$\bullet$
{DC, PO, TPP}	$\bullet$	$\bullet$	$\bullet$
{EC, PO, TPP}	$\bullet$	$\bullet$	$\bullet$
{DC, EC, PO, TPP}	$\bullet$	$\bullet$	$\bullet$
{NTPP}	$\bullet$	$\bullet$	$\bullet$
{DC, NTPP}	$\bullet$	$\bullet$	$\bullet$
{EC, NTPP}	$\bullet$		$\bullet$
{DC, EC, NTPP}	$\bullet$		$\bullet$
{PO, NTPP}	$\bullet$	$\bullet$	$\bullet$
{DC, PO, NTPP}	$\bullet$	$\bullet$	$\bullet$
{EC, PO, NTPP}	$\bullet$	$\bullet$	$\bullet$
{DC, EC, PO, NTPP}	$\bullet$	$\bullet$	$\bullet$
{TPP, NTPP}	$\bullet$	$\bullet$	$\bullet$
{DC, TPP, NTPP}	$\bullet$	$\bullet$	$\bullet$
{EC, TPP, NTPP}	$\bullet$		$\bullet$
{DC, EC, TPP, NTPP}	$\bullet$		$\bullet$
{PO, TPP, NTPP}	$\bullet$	$\bullet$	$\bullet$
{DC, PO, TPP, NTPP}	$\bullet$	$\bullet$	$\bullet$
{EC, PO, TPP, NTPP}	$\bullet$	$\bullet$	$\bullet$
{DC, EC, PO, TPP, NTPP}	$\bullet$	$\bullet$	$\bullet$

Relation	$\widehat{\mathcal{H}}_8$	$\mathcal{C}_8$	$\mathcal{Q}_8$
$\{\mathsf{TPP}^{-1}\}$	•	•	•
$\{\mathsf{DC}, \mathsf{TPP}^{-1}\}$	•	•	•
$\{\mathsf{EC}, \mathsf{TPP}^{-1}\}$	•		•
$\{\mathsf{DC}, \mathsf{EC}, \mathsf{TPP}^{-1}\}$	•		•
$\{\mathsf{PO}, \mathsf{TPP}^{-1}\}$	•	•	•
$\{\mathsf{DC}, \mathsf{PO}, \mathsf{TPP}^{-1}\}$	•	•	•
$\{\mathsf{EC}, \mathsf{PO}, \mathsf{TPP}^{-1}\}$	•	•	•
$\{\mathsf{DC}, \mathsf{EC}, \mathsf{PO}, \mathsf{TPP}^{-1}\}$	•	•	•
$\{\mathsf{TPP}, \mathsf{TPP}^{-1}\}$			
$\{\mathsf{DC}, \mathsf{TPP}, \mathsf{TPP}^{-1}\}$			
$\{\mathsf{EC}, \mathsf{TPP}, \mathsf{TPP}^{-1}\}$			
$\{\mathsf{DC}, \mathsf{EC}, \mathsf{TPP}, \mathsf{TPP}^{-1}\}$			
$\{\mathsf{PO}, \mathsf{TPP}, \mathsf{TPP}^{-1}\}$	•	•	•
$\{\mathsf{DC}, \mathsf{PO}, \mathsf{TPP}, \mathsf{TPP}^{-1}\}$	•	•	•
$\{\mathsf{EC}, \mathsf{PO}, \mathsf{TPP}, \mathsf{TPP}^{-1}\}$	•	•	•
$\{\mathsf{DC}, \mathsf{EC}, \mathsf{PO}, \mathsf{TPP}, \mathsf{TPP}^{-1}\}$	•	•	•
$\{\mathsf{NTPP}, \mathsf{TPP}^{-1}\}$			
$\{\mathsf{DC}, \mathsf{NTPP}, \mathsf{TPP}^{-1}\}$			
$\{\mathsf{EC}, \mathsf{NTPP}, \mathsf{TPP}^{-1}\}$			
$\{\mathsf{DC}, \mathsf{EC}, \mathsf{NTPP}, \mathsf{TPP}^{-1}\}$			
$\{\mathsf{PO}, \mathsf{NTPP}, \mathsf{TPP}^{-1}\}$	•	•	•
$\{\mathsf{DC}, \mathsf{PO}, \mathsf{NTPP}, \mathsf{TPP}^{-1}\}$	•	•	•
$\{\mathsf{EC}, \mathsf{PO}, \mathsf{NTPP}, \mathsf{TPP}^{-1}\}$	•	•	•
$\{\mathsf{DC}, \mathsf{EC}, \mathsf{PO}, \mathsf{NTPP}, \mathsf{TPP}^{-1}\}$	•	•	•
$\{\mathsf{TPP}, \mathsf{NTPP}, \mathsf{TPP}^{-1}\}$			
$\{\mathsf{DC}, \mathsf{TPP}, \mathsf{NTPP}, \mathsf{TPP}^{-1}\}$			
$\{\mathsf{EC}, \mathsf{TPP}, \mathsf{NTPP}, \mathsf{TPP}^{-1}\}$			
$\{\mathsf{DC}, \mathsf{EC}, \mathsf{TPP}, \mathsf{NTPP}, \mathsf{TPP}^{-1}\}$			
$\{\mathsf{PO}, \mathsf{TPP}, \mathsf{NTPP}, \mathsf{TPP}^{-1}\}$	•	•	•
$\{\mathsf{DC}, \mathsf{PO}, \mathsf{TPP}, \mathsf{NTPP}, \mathsf{TPP}^{-1}\}$	•	•	•
$\{\mathsf{EC}, \mathsf{PO}, \mathsf{TPP}, \mathsf{NTPP}, \mathsf{TPP}^{-1}\}$	•	•	•
$\{\mathsf{DC}, \mathsf{EC}, \mathsf{PO}, \mathsf{TPP}, \mathsf{NTPP}, \mathsf{TPP}^{-1}\}$	•	•	•
$\{\mathsf{NTPP}^{-1}\}$	•	•	•
$\{\mathsf{DC}, \mathsf{NTPP}^{-1}\}$	•	•	•
$\{\mathsf{EC}, \mathsf{NTPP}^{-1}\}$	•		•
$\{\mathsf{DC}, \mathsf{EC}, \mathsf{NTPP}^{-1}\}$	•		•
$\{\mathsf{PO}, \mathsf{NTPP}^{-1}\}$	•	•	•

Relation	$\widehat{\mathcal{H}}_8$	$\mathcal{C}_8$	$\mathcal{Q}_8$
$\{DC, PO, NTPP^{-1}\}$	•	•	•
$\{EC, PO, NTPP^{-1}\}$	•	•	•
$\{DC, EC, PO, NTPP^{-1}\}$	•	•	•
$\{TPP, NTPP^{-1}\}$			
$\{DC, TPP, NTPP^{-1}\}$			
$\{EC, TPP, NTPP^{-1}\}$			
$\{DC, EC, TPP, NTPP^{-1}\}$			
$\{PO, TPP, NTPP^{-1}\}$	•	•	•
$\{DC, PO, TPP, NTPP^{-1}\}$	•	•	•
$\{EC, PO, TPP, NTPP^{-1}\}$	•	•	•
$\{DC, EC, PO, TPP, NTPP^{-1}\}$	•	•	•
$\{NTPP, NTPP^{-1}\}$			
$\{DC, NTPP, NTPP^{-1}\}$			
$\{EC, NTPP, NTPP^{-1}\}$			
$\{DC, EC, NTPP, NTPP^{-1}\}$			
$\{PO, NTPP, NTPP^{-1}\}$	•	•	•
$\{DC, PO, NTPP, NTPP^{-1}\}$	•	•	•
$\{EC, PO, NTPP, NTPP^{-1}\}$	•	•	•
$\{DC, EC, PO, NTPP, NTPP^{-1}\}$	•	•	•
$\{TPP, NTPP, NTPP^{-1}\}$			
$\{DC, TPP, NTPP, NTPP^{-1}\}$			
$\{EC, TPP, NTPP, NTPP^{-1}\}$			
$\{DC, EC, TPP, NTPP, NTPP^{-1}\}$			
$\{PO, TPP, NTPP, NTPP^{-1}\}$	•	•	•
$\{DC, PO, TPP, NTPP, NTPP^{-1}\}$	•	•	•
$\{EC, PO, TPP, NTPP, NTPP^{-1}\}$	•	•	•
$\{DC, EC, PO, TPP, NTPP, NTPP^{-1}\}$	•	•	•
$\{TPP^{-1}, NTPP^{-1}\}$	•	•	•
$\{DC, TPP^{-1}, NTPP^{-1}\}$	•	•	•
$\{EC, TPP^{-1}, NTPP^{-1}\}$	•		•
$\{DC, EC, TPP^{-1}, NTPP^{-1}\}$	•		•
$\{PO, TPP^{-1}, NTPP^{-1}\}$	•	•	•
$\{DC, PO, TPP^{-1}, NTPP^{-1}\}$	•	•	•
$\{EC, PO, TPP^{-1}, NTPP^{-1}\}$	•	•	•
$\{DC, EC, PO, TPP^{-1}, NTPP^{-1}\}$	•	•	•
$\{TPP, TPP^{-1}, NTPP^{-1}\}$			
$\{DC, TPP, TPP^{-1}, NTPP^{-1}\}$			

Relation	$\widehat{\mathcal{H}}_8$	$\mathcal{C}_8$	$\mathcal{Q}_8$
$\{\mathsf{EC}, \mathsf{TPP}, \mathsf{TPP}^{-1}, \mathsf{NTPP}^{-1}\}$			
$\{\mathsf{DC}, \mathsf{EC}, \mathsf{TPP}, \mathsf{TPP}^{-1}, \mathsf{NTPP}^{-1}\}$			
$\{\mathsf{PO}, \mathsf{TPP}, \mathsf{TPP}^{-1}, \mathsf{NTPP}^{-1}\}$	•	•	•
$\{\mathsf{DC}, \mathsf{PO}, \mathsf{TPP}, \mathsf{TPP}^{-1}, \mathsf{NTPP}^{-1}\}$	•	•	•
$\{\mathsf{EC}, \mathsf{PO}, \mathsf{TPP}, \mathsf{TPP}^{-1}, \mathsf{NTPP}^{-1}\}$	•	•	•
$\{\mathsf{DC}, \mathsf{EC}, \mathsf{PO}, \mathsf{TPP}, \mathsf{TPP}^{-1}, \mathsf{NTPP}^{-1}\}$	•	•	•
$\{\mathsf{NTPP}, \mathsf{TPP}^{-1}, \mathsf{NTPP}^{-1}\}$			
$\{\mathsf{DC}, \mathsf{NTPP}, \mathsf{TPP}^{-1}, \mathsf{NTPP}^{-1}\}$			
$\{\mathsf{EC}, \mathsf{NTPP}, \mathsf{TPP}^{-1}, \mathsf{NTPP}^{-1}\}$			
$\{\mathsf{DC}, \mathsf{EC}, \mathsf{NTPP}, \mathsf{TPP}^{-1}, \mathsf{NTPP}^{-1}\}$			
$\{\mathsf{PO}, \mathsf{NTPP}, \mathsf{TPP}^{-1}, \mathsf{NTPP}^{-1}\}$	•	•	•
$\{\mathsf{DC}, \mathsf{PO}, \mathsf{NTPP}, \mathsf{TPP}^{-1}, \mathsf{NTPP}^{-1}\}$	•	•	•
$\{\mathsf{EC}, \mathsf{PO}, \mathsf{NTPP}, \mathsf{TPP}^{-1}, \mathsf{NTPP}^{-1}\}$	•	•	•
$\{\mathsf{DC}, \mathsf{EC}, \mathsf{PO}, \mathsf{NTPP}, \mathsf{TPP}^{-1}, \mathsf{NTPP}^{-1}\}$	•	•	•
$\{\mathsf{TPP}, \mathsf{NTPP}, \mathsf{TPP}^{-1}, \mathsf{NTPP}^{-1}\}$			
$\{\mathsf{DC}, \mathsf{TPP}, \mathsf{NTPP}, \mathsf{TPP}^{-1}, \mathsf{NTPP}^{-1}\}$			
$\{\mathsf{EC}, \mathsf{TPP}, \mathsf{NTPP}, \mathsf{TPP}^{-1}, \mathsf{NTPP}^{-1}\}$			
$\{\mathsf{DC}, \mathsf{EC}, \mathsf{TPP}, \mathsf{NTPP}, \mathsf{TPP}^{-1}, \mathsf{NTPP}^{-1}\}$			
$\{\mathsf{PO}, \mathsf{TPP}, \mathsf{NTPP}, \mathsf{TPP}^{-1}, \mathsf{NTPP}^{-1}\}$	•	•	•
$\{\mathsf{DC}, \mathsf{PO}, \mathsf{TPP}, \mathsf{NTPP}, \mathsf{TPP}^{-1}, \mathsf{NTPP}^{-1}\}$	•	•	•
$\{\mathsf{EC}, \mathsf{PO}, \mathsf{TPP}, \mathsf{NTPP}, \mathsf{TPP}^{-1}, \mathsf{NTPP}^{-1}\}$	•	•	•
$\{\mathsf{DC}, \mathsf{EC}, \mathsf{PO}, \mathsf{TPP}, \mathsf{NTPP}, \mathsf{TPP}^{-1}, \mathsf{NTPP}^{-1}\}$	•	•	•
$\{\mathsf{EQ}\}$	•	•	•
$\{\mathsf{DC}, \mathsf{EQ}\}$	•	•	•
$\{\mathsf{EC}, \mathsf{EQ}\}$	•		•
$\{\mathsf{DC}, \mathsf{EC}, \mathsf{EQ}\}$	•		•
$\{\mathsf{PO}, \mathsf{EQ}\}$	•	•	•
$\{\mathsf{DC}, \mathsf{PO}, \mathsf{EQ}\}$	•	•	•
$\{\mathsf{EC}, \mathsf{PO}, \mathsf{EQ}\}$	•	•	•
$\{\mathsf{DC}, \mathsf{EC}, \mathsf{PO}, \mathsf{EQ}\}$	•	•	•
$\{\mathsf{TPP}, \mathsf{EQ}\}$	•	•	
$\{\mathsf{DC}, \mathsf{TPP}, \mathsf{EQ}\}$	•	•	
$\{\mathsf{EC}, \mathsf{TPP}, \mathsf{EQ}\}$	•		
$\{\mathsf{DC}, \mathsf{EC}, \mathsf{TPP}, \mathsf{EQ}\}$	•		
$\{\mathsf{PO}, \mathsf{TPP}, \mathsf{EQ}\}$	•	•	•
$\{\mathsf{DC}, \mathsf{PO}, \mathsf{TPP}, \mathsf{EQ}\}$	•	•	•
$\{\mathsf{EC}, \mathsf{PO}, \mathsf{TPP}, \mathsf{EQ}\}$	•	•	•

Relation	$\widehat{\mathcal{H}}_8$	$\mathcal{C}_8$	$\mathcal{Q}_8$
$\{DC, EC, PO, TPP, EQ\}$	•	•	•
$\{NTPP, EQ\}$		•	
$\{DC, NTPP, EQ\}$		•	
$\{EC, NTPP, EQ\}$			
$\{DC, EC, NTPP, EQ\}$			
$\{PO, NTPP, EQ\}$		•	•
$\{DC, PO, NTPP, EQ\}$		•	•
$\{EC, PO, NTPP, EQ\}$		•	•
$\{DC, EC, PO, NTPP, EQ\}$		•	•
$\{TPP, NTPP, EQ\}$	•	•	
$\{DC, TPP, NTPP, EQ\}$	•	•	
$\{EC, TPP, NTPP, EQ\}$	•		
$\{DC, EC, TPP, NTPP, EQ\}$	•		
$\{PO, TPP, NTPP, EQ\}$	•	•	•
$\{DC, PO, TPP, NTPP, EQ\}$	•	•	•
$\{EC, PO, TPP, NTPP, EQ\}$	•	•	•
$\{DC, EC, PO, TPP, NTPP, EQ\}$	•	•	•
$\{TPP^{-1}, EQ\}$	•	•	
$\{DC, TPP^{-1}, EQ\}$	•	•	
$\{EC, TPP^{-1}, EQ\}$	•		
$\{DC, EC, TPP^{-1}, EQ\}$	•		
$\{PO, TPP^{-1}, EQ\}$	•	•	•
$\{DC, PO, TPP^{-1}, EQ\}$	•	•	•
$\{EC, PO, TPP^{-1}, EQ\}$	•	•	•
$\{DC, EC, PO, TPP^{-1}, EQ\}$	•	•	•
$\{TPP, TPP^{-1}, EQ\}$			
$\{DC, TPP, TPP^{-1}, EQ\}$			
$\{EC, TPP, TPP^{-1}, EQ\}$			
$\{DC, EC, TPP, TPP^{-1}, EQ\}$			
$\{PO, TPP, TPP^{-1}, EQ\}$	•	•	•
$\{DC, PO, TPP, TPP^{-1}, EQ\}$	•	•	•
$\{EC, PO, TPP, TPP^{-1}, EQ\}$	•	•	•
$\{DC, EC, PO, TPP, TPP^{-1}, EQ\}$	•	•	•
$\{NTPP, TPP^{-1}, EQ\}$			
$\{DC, NTPP, TPP^{-1}, EQ\}$			
$\{EC, NTPP, TPP^{-1}, EQ\}$			
$\{DC, EC, NTPP, TPP^{-1}, EQ\}$			
$\{PO, NTPP, TPP^{-1}, EQ\}$		•	•

Relation	$\widehat{\mathcal{H}}_8$	$\mathcal{C}_8$	$\mathcal{Q}_8$
$\{DC, PO, NTPP, TPP^{-1}, EQ\}$		•	•
$\{EC, PO, NTPP, TPP^{-1}, EQ\}$		•	•
$\{DC, EC, PO, NTPP, TPP^{-1}, EQ\}$		•	•
$\{TPP, NTPP, TPP^{-1}, EQ\}$			
$\{DC, TPP, NTPP, TPP^{-1}, EQ\}$			
$\{EC, TPP, NTPP, TPP^{-1}, EQ\}$			
$\{DC, EC, TPP, NTPP, TPP^{-1}, EQ\}$			
$\{PO, TPP, NTPP, TPP^{-1}, EQ\}$	•	•	•
$\{DC, PO, TPP, NTPP, TPP^{-1}, EQ\}$	•	•	•
$\{EC, PO, TPP, NTPP, TPP^{-1}, EQ\}$	•	•	•
$\{DC, EC, PO, TPP, NTPP, TPP^{-1}, EQ\}$	•	•	•
$\{NTPP^{-1}, EQ\}$		•	
$\{DC, NTPP^{-1}, EQ\}$		•	
$\{EC, NTPP^{-1}, EQ\}$			
$\{DC, EC, NTPP^{-1}, EQ\}$			
$\{PO, NTPP^{-1}, EQ\}$		•	•
$\{DC, PO, NTPP^{-1}, EQ\}$		•	•
$\{EC, PO, NTPP^{-1}, EQ\}$		•	•
$\{DC, EC, PO, NTPP^{-1}, EQ\}$		•	•
$\{TPP, NTPP^{-1}, EQ\}$			
$\{DC, TPP, NTPP^{-1}, EQ\}$			
$\{EC, TPP, NTPP^{-1}, EQ\}$			
$\{DC, EC, TPP, NTPP^{-1}, EQ\}$			
$\{PO, TPP, NTPP^{-1}, EQ\}$		•	•
$\{DC, PO, TPP, NTPP^{-1}, EQ\}$		•	•
$\{EC, PO, TPP, NTPP^{-1}, EQ\}$		•	•
$\{DC, EC, PO, TPP, NTPP^{-1}, EQ\}$		•	•
$\{NTPP, NTPP^{-1}, EQ\}$			
$\{DC, NTPP, NTPP^{-1}, EQ\}$			
$\{EC, NTPP, NTPP^{-1}, EQ\}$			
$\{DC, EC, NTPP, NTPP^{-1}, EQ\}$			
$\{PO, NTPP, NTPP^{-1}, EQ\}$		•	•
$\{DC, PO, NTPP, NTPP^{-1}, EQ\}$		•	•
$\{EC, PO, NTPP, NTPP^{-1}, EQ\}$		•	•
$\{DC, EC, PO, NTPP, NTPP^{-1}, EQ\}$		•	•
$\{TPP, NTPP, NTPP^{-1}, EQ\}$			
$\{DC, TPP, NTPP, NTPP^{-1}, EQ\}$			

Relation	$\widehat{\mathcal{H}}_8$	$\mathcal{C}_8$	$\mathcal{Q}_8$
$\{EC, TPP, NTPP, NTPP^{-1}, EQ\}$			
$\{DC, EC, TPP, NTPP, NTPP^{-1}, EQ\}$			
$\{PO, TPP, NTPP, NTPP^{-1}, EQ\}$		•	•
$\{DC, PO, TPP, NTPP, NTPP^{-1}, EQ\}$		•	•
$\{EC, PO, TPP, NTPP, NTPP^{-1}, EQ\}$		•	•
$\{DC, EC, PO, TPP, NTPP, NTPP^{-1}, EQ\}$		•	•
$\{TPP^{-1}, NTPP^{-1}, EQ\}$	•	•	
$\{DC, TPP^{-1}, NTPP^{-1}, EQ\}$	•	•	
$\{EC, TPP^{-1}, NTPP^{-1}, EQ\}$	•		
$\{DC, EC, TPP^{-1}, NTPP^{-1}, EQ\}$	•		
$\{PO, TPP^{-1}, NTPP^{-1}, EQ\}$	•	•	•
$\{DC, PO, TPP^{-1}, NTPP^{-1}, EQ\}$	•	•	•
$\{EC, PO, TPP^{-1}, NTPP^{-1}, EQ\}$	•	•	•
$\{DC, EC, PO, TPP^{-1}, NTPP^{-1}, EQ\}$	•	•	•
$\{TPP, TPP^{-1}, NTPP^{-1}, EQ\}$			
$\{DC, TPP, TPP^{-1}, NTPP^{-1}, EQ\}$			
$\{EC, TPP, TPP^{-1}, NTPP^{-1}, EQ\}$			
$\{DC, EC, TPP, TPP^{-1}, NTPP^{-1}, EQ\}$			
$\{PO, TPP, TPP^{-1}, NTPP^{-1}, EQ\}$	•	•	•
$\{DC, PO, TPP, TPP^{-1}, NTPP^{-1}, EQ\}$	•	•	•
$\{EC, PO, TPP, TPP^{-1}, NTPP^{-1}, EQ\}$	•	•	•
$\{DC, EC, PO, TPP, TPP^{-1}, NTPP^{-1}, EQ\}$	•	•	•
$\{NTPP, TPP^{-1}, NTPP^{-1}, EQ\}$			
$\{DC, NTPP, TPP^{-1}, NTPP^{-1}, EQ\}$			
$\{EC, NTPP, TPP^{-1}, NTPP^{-1}, EQ\}$			
$\{DC, EC, NTPP, TPP^{-1}, NTPP^{-1}, EQ\}$			
$\{PO, NTPP, TPP^{-1}, NTPP^{-1}, EQ\}$		•	•
$\{DC, PO, NTPP, TPP^{-1}, NTPP^{-1}, EQ\}$		•	•
$\{EC, PO, NTPP, TPP^{-1}, NTPP^{-1}, EQ\}$		•	•
$\{DC, EC, PO, NTPP, TPP^{-1}, NTPP^{-1}, EQ\}$		•	•
$\{TPP, NTPP, TPP^{-1}, NTPP^{-1}, EQ\}$			
$\{DC, TPP, NTPP, TPP^{-1}, NTPP^{-1}, EQ\}$			
$\{EC, TPP, NTPP, TPP^{-1}, NTPP^{-1}, EQ\}$			
$\{DC, EC, TPP, NTPP, TPP^{-1}, NTPP^{-1}, EQ\}$			
$\{PO, TPP, NTPP, TPP^{-1}, NTPP^{-1}, EQ\}$	•	•	•
$\{DC, PO, TPP, NTPP, TPP^{-1}, NTPP^{-1}, EQ\}$	•	•	•
$\{EC, PO, TPP, NTPP, TPP^{-1}, NTPP^{-1}, EQ\}$	•	•	•
$\{DC, EC, PO, TPP, NTPP, TPP^{-1}, NTPP^{-1}, EQ\}$	•	•	•

References

1. *Proceedings of the 10th National Conference of the American Association for Artificial Intelligence*, San Jose, CA, July 1992. MIT Press.
2. D. Achlioptas, L. Kirousis, E. Kranakis, D. Krizanc, M. Molloy, and Y. Stamatiou. Random constraint satisfaction: a more accurate picture. In *3rd Conference on the Principles and Practice of Constraint Programming (CP'97)*, volume 1330 of *LNCS*, pages 107–120. Springer, 1997.
3. L.C. Aiello, J. Doyle, and S.C. Shapiro, editors. *Principles of Knowledge Representation and Reasoning: Proceedings of the 5th International Conference*, Cambridge, MA, November 1996. Morgan Kaufmann.
4. James F. Allen. Maintaining knowledge about temporal intervals. *Communications of the ACM*, 26(11):832–843, November 1983.
5. Nicholas Asher and Laure Vieu. Towards a geometry of common sense: A semantics and a complete axiomatization of mereotopology. In IJCAI-95 [94], pages 846–852.
6. Philippe Balbiani, Jean-François Condotta, and Luis Farinas del Cerro. A model for reasoning about bidimensional temporal relations. In Cohn et al. [25], pages 124–130.
7. Philippe Balbiani, Jean-François Condotta, and Luis Farinas del Cerro. A new tractable subclass of the rectangle algebra. In IJCAI-99 [96], pages 442–447.
8. Philippe Balbiani, Jean-François Condotta, and Luis Farinas del Cerro. A tractable subclass of the block algebra: constraint propagation and preconvex relations. In *Proccedings of the 9th Portuguese Conference on Artificial Intelligence*, 1999.
9. John D. Baum. *Elements of Point Set Topology*. Dover, 1991.
10. Brandon Bennett. Spatial reasoning with propositional logic. In J. Doyle, E. Sandewall, and P. Torasso, editors, *Principles of Knowledge Representation and Reasoning: Proceedings of the 4th International Conference*, pages 51–62, Bonn, Germany, May 1994. Morgan Kaufmann.
11. Brandon Bennett. Carving up space: steps towards construction of an absolutely complete theory of spatial regions. In *Logics in Artificial Intelligence, Proceedings of JELIA'96*, volume 1126 of *Lecture Notes in Computer Science*, pages 337–353, 1996.
12. Brandon Bennett. Modal logics for qualitative spatial reasoning. *Bulletin of the IGPL*, 4(1):23 – 45, 1996.
13. Brandon Bennett. *Logical Representations for Automated Reasoning about Spatial Relationships*. PhD thesis, School of Computer Studies, The University of Leeds, 1997.
14. Brandon Bennett, Anthony G. Cohn, and Amar Isli. Combining multiple representations in a spatial reasoning system. In *Proceedings of the 9th IEEE International Conference on Tools with Artificial Intelligence (ICTAI'97)*, pages 314–322, 1997.

15. Brandon Bennett, Amar Isli, and Anthony G. Cohn. When does a composition table provide a complete and tractable proof procedure for a relational constraint language? In *Proceedings of the IJCAI-97 workshop on Spatial and Temporal Reasoning*, Nagoya, Japan, 1997.

16. Loredana Biacino and Giangiacomo Gerla. Connection structures. *Notre Dame Journal of Formal Logic*, 32(2):242–247, 1991.

17. Daniel G. Bobrow. Natural language input for a computer problem solving system. In M. Minsky, editor, *Semantic Information Processing*. MIT Press, Cambridge, MA, 1968.

18. Stefano Borgo, Nicola Guarino, and Claudio Masolo. A pointless theory of space based on strong connection and congruence. In Aiello et al. [3], pages 220–229.

19. Eugene Charniak. Carps, a program which solves calculus word problems. Technical report, MIT, 1968. Project MAC TR-51.

20. Peter Cheeseman, Bob Kanefsky, and William M. Taylor. Where the *really* hard problems are. In *Proceedings of the 12th International Joint Conference on Artificial Intelligence*, pages 331–337, Sydney, Australia, August 1991. Morgan Kaufmann.

21. Brian F. Chellas. *Modal Logic: An Introduction*. Cambridge University Press, Cambridge, UK, 1980.

22. Bowman L. Clarke. A calculus of individuals based on connection. *Notre Dame Journal of Formal Logic*, 22(3):204–218, 1981.

23. Bowman L. Clarke. Individuals and points. *Notre Dame Journal of Formal Logic*, 26(1):61–75, 1985.

24. Eliseo Clementini, Paolino di Felice, and Daniel Hernandez. Qualitative representation of positional information. *ai*, 95(2):317–356, 1997.

25. A.G. Cohn, L. Schubert, and S.C. Shapiro, editors. *Principles of Knowledge Representation and Reasoning: Proceedings of the 6th International Conference*, Trento, Italy, June 1998.

26. Anthony G. Cohn. Qualitative spatial representation and reasoning techniques. In G. Brewka, C. Habel, and B. Nebel, editors, *KI-97: Advances in Artificial Intelligence*, volume 1303 of *Lecture Notes in Computer Science*, pages 1–30, Freiburg, Germany, 1997. Springer-Verlag.

27. Anthony G. Cohn. Qualitative spatial representations. In *Working notes of IJCAI'99 workshop on Adaptive Spatial Representations of Dynamic Environments*, pages 33–52, 1999.

28. Anthony G. Cohn, Brandon Bennett, John Gooday, and Nicholas M. Gotts. Representing and reasoning with qualitative spatial relations about regions. In O. Stock, editor, *Spatial and Temporal Reasoning*, pages 97–134. Kluwer, Dordrecht, Holland, 1997.

29. Anthony G. Cohn and Achille C. Varzi. Connection relations in mereotopology. In ECAI-98 [43], pages 150–154.

30. Anthony G. Cohn and Achille C. Varzi. Modes of connection. In *Spatial Information Theory: Cognitive and Computational Foundations of Geographic Information Science*, pages 299–314, 1999.

31. Stephen A. Cook. The complexity of theorem-proving procedures. In *Proc. 3rd Ann. ACM Symp. on Theory of Computing*, pages 151–158, New York, 1971. Association for Computing Machinery.

32. Thomas H. Cormen, Charles E. Leiserson, and Ronald L. Rivest. *Introduction to Algorithms*. MIT Press, 1990.

33. Zhan Cui, Anthony G. Cohn, and David A. Randell. Qualitative simulation based on a logical formalism of space and time. In AAAI-92 [1], pages 679 – 684.

34. Ernest Davis. *Representation of Commonsense Knowledge*. Morgan Kaufmann, San Mateo, CA, 1990.
35. Ernest Davis, Nicholas M. Gotts, and Anthony G. Cohn. Constraint networks of topological relations and convexity. *CONSTRAINTS*, 4(3):241–280, 1999.
36. Martin Davis. First-order logic. In Gabbay et al. [59], pages 31–67.
37. Johan de Kleer. Qualitative and quantitative knowledge in classical mechanics. Technical Report Artificial Intelligence Laboratory TR-352, M.I.T., 1975.
38. Rina Dechter. From local to global consistency. *Artificial Intelligence*, 55:87–108, 1992.
39. Christoph Dornheim. Undecidability of plane polygonal mereotopology. In Cohn et al. [25].
40. William F. Dowling and Jean H. Gallier. Linear time algorithms for testing the satisfiability of propositional Horn formula. *The Journal of Logic Programming*, 3:267–284, 1984.
41. Thomas Drakengren and Peter Jonsson. Twenty-one large tractable subclasses of Allen's algebra. *Artificial Intelligence*, 93(1-2):297–319, 1997.
42. Thomas Drakengren and Peter Jonsson. A complete classification of tractability in Allen's algebra relative to subsets of basic relations. *Artificial Intelligence*, 106(2):205–219, 1998.
43. *Proceedings of the 13th European Conference on Artificial Intelligence*, Amsterdam, The Netherlands, 1998. Wiley.
44. Max J. Egenhofer. Reasoning about binary topological relations. In O. Günther and H.-J. Schek, editors, *Proceedings of the Second Symposium on Large Spatial Databases, SSD'91*, volume 525 of *Lecture Notes in Computer Science*, pages 143–160. Springer-Verlag, Berlin, Heidelberg, New York, 1991.
45. Max J. Egenhofer, Eliseo Clementini, and Paolino Di Felice. Topological relations between regions with holes. *International Journal of Geographical Information Systems*, 8(2):129–144, 1994.
46. Max J. Egenhofer and Robert D. Franzosa. On the equivalence of topological relations. *International Journal of Geographical Information Systems*, 8(6):133–152, 1994.
47. Max J. Egenhofer and Reginald G. Golledge, editors. *Spatial and Temporal Reasoning in Geographic Information Systems*. Oxford University Press, Oxford, UK, 1998.
48. Max J. Egenhofer and Jayant Sharma. Assessing the consistency of complete and incomplete topological information. *Geographical Systems*, 1(1):47–68, 1993.
49. Maria T. Escrig. *Qualitative Spatial Reasoning: theory and practice, Application to Robot Navigation*. IOS Press, 1998.
50. Melvin C. Fitting. *Proof Methods for Modal and Intuitionistic Logics*. Reidel, Dordrecht, Holland, 1983.
51. Melvin C. Fitting. Basic modal logic. In Gabbay et al. [59], pages 365–448.
52. Kenneth D. Forbus. Spatial and qualitative aspects of reasoning about motion. In *Proceedings of the 1st National Conference of the American Association for Artificial Intelligence*, Stanford, CA, August 1980.
53. Kenneth D. Forbus. Qualitative reasoning about physical processes. In *Proceedings of the 7th International Joint Conference on Artificial Intelligence*, Vancouver, Canada, August 1981.
54. Kenneth D. Forbus. Qualitative physics: Past, present, and future. In D. S. Weld and J. de Kleer, editors, *Readings in Qualitative Reasoning about Physical Systems*, pages 11–39. Morgan Kaufmann, San Mateo, CA, 1989.

55. Kenneth D. Forbus, Paul Nielsen, and Boi Faltings. Qualitative kinematics: A framework. In J. McDermott, editor, *Proceedings of the 10th International Joint Conference on Artificial Intelligence*, Milan, Italy, August 1987. Morgan Kaufmann.

56. Andrew U. Frank. Qualitative spatial reasoning about cardinal directions. In *Proceedings of the 7th Austrian Conference on Artificial Intelligence*, pages 157–167, 1991.

57. C. Freksa, C. Habel, and K. F. Wender, editors. *Spatial cognition. An interdisciplinary approach to representing and processing spatial knowledge*, volume 1404 of *Lecture Notes in Artificial Intelligence*. Springer-Verlag, Berlin, Heidelberg, New York, 1998.

58. Christian Freksa. Using orientation information for qualitative spatial reasoning. In U. Formentini A.U. Frank, I. Campari, editor, *Theories and Methods of Spatio-Temporal Reasoning in Geographic Space*, volume 639 of *Lecture Notes in Computer Science*. Springer-Verlag, Berlin, Heidelberg, New York, 1992.

59. D. M. Gabbay, C. J. Hogger, and J. A. Robinson, editors. *Handbook of Logic in Artificial Intelligence and Logic Programming – Vol. 1: Logical Foundations*. Oxford, Clarendon Press, 1993.

60. Jean H. Gallier. *Logic for Computer Science*. Harper and Row, New York, 1986.

61. Antony Galton. Mereotopology of discrete space. In C. Freksa and D.M. Mark, editors, *Spatial information theory: Cognitive and computational foundations of geographic information science*, volume 1661 of *LNCS*, pages 251–266, springort, 1999. spring.

62. Michael R. Garey and David S. Johnson. *Computers and Intractability—A Guide to the Theory of NP-Completeness*. Freeman, San Francisco, CA, 1979.

63. Michael R. Genesereth and Nils J. Nilsson. *Logical Foundations of Artificial Intelligence*. Morgan Kaufmann, Los Altos, CA, 1987.

64. I. Gent, E. MacIntyre, P. Prosser, B. Smith, and T. Walsh. Random constraint satisfaction: Flaws and structure. Technical report, APES research group, http://www.cs.strath.edu.uk/ apes/apesreports.html, 1998.

65. I. Gent, E. MacIntyre, P. Prosser, and T. Walsh. The constrainedness of search. In *Proceedings of the 13th National Conference on AI (AAAI'96)*, pages 246–252, 1996.

66. Ian P. Gent and Toby Walsh. The satisfiability constraint gap. *Artificial Intelligence*, 81(1–2):59–80, 1996.

67. Alfonso Gerevini and Matteo Cristani. On finding a solution in temporal constraint satisfaction problems. In IJCAI-97 [95], pages 1460–1465.

68. Alfonso Gerevini and Jochen Renz. Combining topological and qualitative size constraints for spatial reasoning. In *Proceedings of the 4th International Conference on Principles and Practice of Constraint Programming*, Pisa, Italy, 1998.

69. Martin C. Golumbic and Ron Shamir. Complexity and algorithms for reasoning about time: A graph-theoretic approach. *Journal of the Association for Computing Machinery*, 40(5):1128–1133, November 1993.

70. John M. Gooday and Anthony G. Cohn. Using spatial logic to describe visual programming languages. *Artificial Intelligence Review*, 10:171–186, 1996.

71. Nicholas M. Gotts. How far can we C? defining a 'doughnut' using connection alone. In E. Sandewall J. Doyle and P. Torasso, editors, *Principles of Knowledge Representation and Reasoning: Proceedings of the 4th International Conference (KR94)*, pages 246–257, San Francisco, 1994. Morgan Kaufmann.

72. Nicholas M. Gotts. An axiomatic approach to topology for spatial information systems. Technical Report 96-25, University of Leeds, School of Computer Studies, 1996.
73. Nicholas M. Gotts. Topology from a single primitive relation: Defining topological properties and relations in terms of connection. Technical Report 96-23, University of Leeds, School of Computer Studies, 1996.
74. Nicholas M. Gotts. Using the RCC formalism to describe the topology of spherical regions. Technical Report 96-24, University of Leeds, School of Computer Studies, 1996.
75. Nicholas M. Gotts, John M. Gooday, and Anthony G. Cohn. A connection based approach to commonsense topological description and reasoning. *The Monist*, 79(1):51–75, 1996.
76. Roop Goyal and Max J. Egenhofer. Cardinal directions between extended spatial objects. *IEEE Transactions on Knowledge and Data Engineering*, 2000. to appear.
77. Michelangelo Grigni, Dimitris Papadias, and Christos Papadimitriou. Topological inference. In IJCAI-95 [94], pages 901–906.
78. Andrzej Grzegorczyk. Undecidability of some topological theories. *Fundamenta Mathematicae*, 38:137–152, 1951.
79. Hans Guesgen. Spatial reasoning based on Allen's temporal logic. Technical Report TR-89-049, ICSI, Berkeley, CA, 1989.
80. Jens-Steffen Gutmann, Wolfgang Hatzack, Immanuel Herrmann, Bernhard Nebel, Frank Rittinger, Augustinus Topor, and Thilo Weigel. The CS Freiburg team: Playing robotic soccer on an explicit world model. *The AI Magazine*, 1(21):37–46, 2000.
81. Volker Haarslev, Carsten Lutz, and Ralf Möller. Foundations of spatioterminological reasoning with description logics. In Cohn et al. [25], pages 112–123.
82. Patrick J. Hayes. The naive physics manifesto. In D. Michie, editor, *Expert Systems in the microelectronic age*. Edinburgh University Press, Edinburgh, UK, 1978.
83. Patrick J. Hayes. Naive physics I: Ontology for liquids. In *Formal Theories of the Commonsense World* [90], pages 71–108.
84. Patrick J. Hayes. The second naive physics manifesto. In *Formal Theories of the Commonsense World* [90], pages 1–36.
85. William G. Hayward and Michael J. Tarr. Spatial language and spatial representation. *Cognition*, 55:39–84, 1995.
86. Lawrence J. Henschen and Larry Wos. Unit refutations and Horn sets. *Journal of the Association for Computing Machinery*, 21:590–605, 1974.
87. Daniel Hernàndez. *Qualitative Representation of Spatial Knowledge*, volume 804 of *Lecture Notes in Artificial Intelligence*. Springer-Verlag, Berlin, Heidelberg, New York, 1994.
88. Daniel Hernàndez, Eliseo Clementini, and Paolino di Felice. Qualitative distances. In *Spatial Information Theory: A Theoretical basis for GIS*, volume 988 of *LNCS*, pages 45–58, Berlin, Heidelberg, New York, 1995. Springer-Verlag.
89. Robin Hirsch. A finite relation algebra with undecidable network satisfaction problem. *Bulletin of the IGPL*, 1999.
90. Jerry R. Hobbs and Robert C. Moore. *Formal Theories of the Commonsense World*. Ablex, Norwood, NJ, 1985.
91. Tad Hogg, Bernardo A. Huberman, and Colin P. Williams (eds). Special volume on frontiers in problem solving: Phase transistions and complexity. *Artificial Intelligence*, 81(1–2), 1996.

92. B. Huberman, R. Lukose, and T. Hogg. An economics approach to hard computational problems. *Science*, 275:51–54, 1997.

93. George E. Hughes and Max J. Cresswell. *An Introduction to Modal Logic*. University Press, 1968.

94. *Proceedings of the 14th International Joint Conference on Artificial Intelligence*, Montreal, Canada, August 1995.

95. *Proceedings of the 15th International Joint Conference on Artificial Intelligence*, Nagoya, Japan, August 1997.

96. *Proceedings of the 16th International Joint Conference on Artificial Intelligence*, Stockholm, Sweden, August 1999.

97. Amar Isli and Reinhard Moratz. Qualitative spatial representation and reasoning: Algebraic models for relative position. Technical Report 284, Universität Hamburg, Fachbereich Informatik, 1999.

98. Bernd Jähne, Horst Haußecker, and Peter Geißler, editors. *Handbook Of Computer Vision and Applications*. Academic Press, Boston, USA, 1999.

99. Nicod Jean. *La géometrie dans le monde sensible*. PhD thesis, Sorbonne, Paris, 1924. English translation in J. Nicod, Geopmetry and Induction, Routledge and Kegan Paul, London, 1970.

100. David S. Johnson. A catalog of complexity classes. In J. van Leeuwen, editor, *Handbook of Theoretical Computer Science, Vol. A*, pages 67–161. MIT Press, 1990.

101. Peter Jonsson and Thomas Drakengren. A complete classification of tractability in RCC-5. *Journal of Artificial Intelligence Research*, 6:211–221, 1997.

102. Markus Knauff, Reinhold Rauh, and Jochen Renz. A cognitive assessment of topological spatial relations: Results from an empirical investigation. In *Proceedings of the 3rd International Conference on Spatial Information Theory (COSIT'97)*, volume 1329 of *Lecture Notes in Computer Science*, pages 193–206, 1997.

103. Markus Knauff, Reinhold Rauh, and Christoph Schlieder. Preferred mental models in qualitative spatial reasoning: A cognitive assessment of Allen's calculus. In *Proceedings of the Seventeenth Annual Conference of the Cognitive Science Society*, pages 200–205, Mahwah, NJ, 1995. Lawrence Erlbaum Associates.

104. Lloyd K. Komatsu. Recent views of conceptual structure. *Psychological Bulletin*, 112:500–526, 1992.

105. Jan Kratochvíl. String graphs II: recognizing string graphs is NP-hard. *Journal of Combinatorial Theory, Series B*, 52:67–78, 1991.

106. Jan Kratochvíl and Jiří Matoušek. String graphs requiring exponential representations. *Journal of Combinatorial Theory, Series B*, 53:1–4, 1991.

107. A. Krokhin, P. Jeavons, and P. Jonsson. A complete classification of complexity in Allen's algebra in the presence of a non-trivial basic relation. In *Proceedings of the 17th International Joint Conference on Artificial Intelligence*, pages 83–88, Seattle, WA, August 2001.

108. Benjamin Kuipers. *Qualitative Reasoning : Modeling and Simulation With Incomplete Knowledge*. MIT Press, Cambridge, MA, 1994.

109. Peter B. Ladkin and Roger Maddux. On binary constraint networks. Technical Report KES.U.88.8, Kestrel Institute, Palo Alto, CA, 1988.

110. Peter B. Ladkin and Roger Maddux. On binary constraint problems. *Journal of the Association for Computing Machinery*, 41(3):435–469, May 1994.

111. Peter B. Ladkin and Alexander Reinefeld. Effective solution of qualitative interval constraint problems. *Artificial Intelligence*, 57(1):105–124, September 1992.

112. Peter B. Ladkin and Alexander Reinefeld. Fast algebraic methods for interval constraint problems. *Annals of Mathematics and Artificial Intelligence*, 19:383–411, 1997.
113. Jean-Claude Latombe. *Robot Motion Planning*. Kluwer, Dordrecht, Holland, 1991.
114. Oliver Lemon. Semantical foundations of spatial logics. In Aiello et al. [3], pages 212–219.
115. Gerard Ligozat. A new proof of tractability for Ord-Horn relations. In *Proceedings of the 13th National Conference of the American Association for Artificial Intelligence*, pages 715–720, Portland, OR, August 1996. MIT Press.
116. Gerard Ligozat. Reasoning about cardinal directions. *Journal of Visual Languages and Computing*, 9:23–44, 1998.
117. Wolfgang Maaß, Peter Wazinski, and Gerd Herzog. VITRA GUIDE: Multimodal route descriptions for computer assisted vehicle navigation. In *Proc. of the Sixth Int. Conf. on Industrial and Engineering Applications of Artificial Intelligence and Expert Systems IEA/AIE-93*, pages 144–147, Edinburgh, Scotland, 1993.
118. Alan K. Mackworth. Consistency in networks of relations. *Artificial Intelligence*, 8:99–118, 1977.
119. Alan K. Mackworth and Eugene C. Freuder. The complexity of some polynomial network consistency algorithms for constraint satisfaction problems. *Artificial Intelligence*, 25:65–73, 1985.
120. David M. Mark, David Comas, Max J. Egenhofer, Scott M. Freundschuh, Michael D. Gould, and Joan Nunes. Evaluation and refining computational models of spatial relations through cross-linguistic human-subjects testing. In A. U. Frank and W. Kuhn, editors, *Spatial Information Theory*, volume 988 of *Lecture Notes in Computer Science*, pages 553–568, Berlin, Heidelberg, New York, 1995. Springer-Verlag.
121. John McCarthy. Programs with common sense. In M. Minsky, editor, *Semantic Information Processing*, pages 403 – 418. MIT Press, Cambridge, MA, 1968.
122. George A. Miller. The magical number seven, plus or minus two: Some limits on our capacity for processing information. *Psychological Review*, 63:81–97, 1956.
123. Ugo Montanari. Networks of constraints: fundamental properties and applications to picture processing. *Information Science*, 7:95–132, 1974.
124. Daniel R. Montello. Scale and multiple psychologies of space. In A.U. Frank and I. Campari, editors, *Spatial information Theory: A theoretical basis for GIS*, volume 716 of *LNCS*, pages 312–321, springort, 1993. spring.
125. James R. Munkres. *Topology; A First Course*. Prentice-Hall, Englewood Cliffs, NJ, 1974.
126. Bernhard Nebel. Computational properties of qualitative spatial reasoning: First results. In I. Wachsmuth, C.-R. Rollinger, and W. Brauer, editors, *KI-95: Advances in Artificial Intelligence*, volume 981 of *Lecture Notes in Artificial Intelligence*, pages 233–244, Bielefeld, Germany, 1995. Springer-Verlag.
127. Bernhard Nebel. Solving hard qualitative temporal reasoning problems: Evaluating the efficiency of using the ORD-Horn class. *CONSTRAINTS*, 3(1):175–190, 1997.
128. Bernhard Nebel and Hans-Jürgen Bürckert. Reasoning about temporal relations: A maximal tractable subclass of Allen's interval algebra. *Journal of the Association for Computing Machinery*, 42(1):43–66, January 1995.

129. Dimitris Papadias and Yannis Theodoridis. Spatial relations,minimum bounding rectangles, and spatial data structures. *International Journal of Geographic Information Systems*, 11(2):111–138, 1997.
130. Christos H. Papadimitriou. *Computational Complexity*. Addison-Wesley, Reading, MA, 1994.
131. Jean Piaget and Bärbel Inhelder. *La représentation de l'espace chez l'enfant*. Presses universitaires de France, Paris, France, 1948.
132. Ian Pratt and Dominik Schoop. A complete axiom system for polygonal mereotopology of the real plane. *Journal of Philosophical Logic*, 27:621–658, 1998.
133. Franco P. Preparata and Michael I. Shamos. *Computational Geometry : An Introduction*. Springer-Verlag, Berlin, Heidelberg, New York, 1991.
134. David A. Randell and Anthony G. Cohn. Modelling topological and metrical properties in physical processes. In R. Brachman, H. J. Levesque, and R. Reiter, editors, *Principles of Knowledge Representation and Reasoning: Proceedings of the 1st International Conference*, pages 55–66, Toronto, ON, May 1989. Morgan Kaufmann.
135. David A. Randell and Anthony G. Cohn. Exploiting lattices in a theory of space and time. *Computers and Mathematics with Applications*, 23(6-9):459–476, 1992.
136. David A. Randell and Anthony G. Cohn. Naive topology: modelling the force pump. In Peter Struss and Boi Faltings, editors, *Advances in Qualitative Physics*, pages 177–192. MIT Press, 1992.
137. David A. Randell, Anthony G. Cohn, and Zhan Cui. Computing transitivity tables: A challenge for automated theorem provers. In *Proceedings of the 11th CADE*. Springer-Verlag, 1992.
138. David A. Randell, Zhan Cui, and Anthony G. Cohn. A spatial logic based on regions and connection. In B. Nebel, W. Swartout, and C. Rich, editors, *Principles of Knowledge Representation and Reasoning: Proceedings of the 3rd International Conference*, pages 165–176, Cambridge, MA, October 1992. Morgan Kaufmann.
139. Jochen Renz. A canonical model of the Region Connection Calculus. In Cohn et al. [25], pages 330 – 341.
140. Jochen Renz. Maximal tractable fragments of the Region Connection Calculus: A complete analysis. In IJCAI-99 [96], pages 448–454.
141. Jochen Renz and Bernhard Nebel. On the complexity of qualitative spatial reasoning: A maximal tractable fragment of the Region Connection Calculus. In IJCAI-97 [95], pages 522–527.
142. Jochen Renz and Bernhard Nebel. Efficient methods for qualitative spatial reasoning. In ECAI-98 [43], pages 562 – 566.
143. Jochen Renz and Bernhard Nebel. On the complexity of qualitative spatial reasoning: A maximal tractable fragment of the Region Connection Calculus. *Artificial Intelligence*, 108(1-2):69–123, 1999.
144. Jochen Renz and Bernhard Nebel. Efficient methods for qualitative spatial reasoning. *Journal of Artificial Intelligence Research*, to appear.
145. Jochen Renz, Reinhold Rauh, and Markus Knauff. Towards cognitive adequacy of topological spatial relations. In C. Freksa, W. Brauer, C. Habel, and K.F. Wender, editors, *Spatial Cognition II—Integrating abstract theories, empirical studies, formal models, and practical applications*, volume 1849 of *LNCS*, pages 184–197. Springer-Verlag, Berlin, Heidelberg, New York, 2000.
146. Lance J. Rips. The current status of research on concept. *Mind and Language*, 10:72–104, 1995.

147. Thomas J. Schaefer. The complexity of satisfiability problems. In *Proc. 10th Ann. ACM Symp. on Theory of Computing*, pages 216–226, New York, 1978. Association for Computing Machinery.

148. Dominik Schoop. *A model-theoretic approach to mereotopology*. PhD thesis, Faculty of Science and Engineering, University of Manchester, 1999.

149. Alexander Scivos and Bernhard Nebel. Double-crossing is harmful: First results on formal properties of a qualitative calculus for navigation. unpublished manuscript, 1999.

150. Bart Selman, Hector J. Levesque, and David Mitchell. A new method for solving hard satisfiability problems. In AAAI-92 [1], pages 440–446.

151. Terence R. Smith and Keith K. Park. Algebraic approach to spatial reasoning. *International Journal of Geographic Information Systems*, 6(3):177–192, 1992.

152. John Stell and Mike Worboys. The algebraic structure of sets of regions. In *Proceedings of the 3rd International Conference on Spatial Information Theory (COSIT'97)*, volume 1329 of *Lecture Notes in Computer Science*, pages 163–174, 1997.

153. Gerhard Strube. The role of cognitive science in knowledge engineering and cognition. In F. Schmalhofer, G. Strube, and T. Wetter, editors, *Contemporary knowledge engineering and cognition*, pages 161–174. Springer-Verlag, 1992.

154. Alfred Tarski. On the calculus of relations. *Journal of Symbolic Logic*, 6:73–89, 1941.

155. Peter van Beek. Reasoning about qualitative temporal information. *Artificial Intelligence*, 58(1-3):297–321, 1992.

156. Peter van Beek and Robin Cohen. Exact and approximate reasoning about temporal relations. *Computational Intelligence*, 6:132–144, 1990.

157. Peter van Beek and Dennis W. Manchak. The design and experimental analysis of algorithms for temporal reasoning. *Journal of Artificial Intelligence Research*, 4:1–18, 1996.

158. Johan van Benthem. Correspondence theory. In D. Gabbay and F. Guenthner, editors, *Handbook of Philosophical Logic, Vol. II*, pages 167–248. Reidel, 1984.

159. Laure Vieu. Spatial representation and reasoning in AI. In O. Stock, editor, *Spatial and Temporal Reasoning*, pages 5–40. Kluwer, Dordrecht, Holland, 1997.

160. Marc B. Vilain and Henry A. Kautz. Constraint propagation algorithms for temporal reasoning. In *Proceedings of the 5th National Conference of the American Association for Artificial Intelligence*, pages 377–382, Philadelphia, PA, August 1986.

161. Marc B. Vilain, Henry A. Kautz, and Peter van Beek. Constraint propagation algorithms for temporal reasoning: A revised report. In D. S. Weld and J. de Kleer, editors, *Readings in Qualitative Reasoning about Physical Systems*, pages 373–381. Morgan Kaufmann, San Mateo, CA, 1989.

162. Constanze Vorwerg and Gert Rickheit. Typicality effects in the categorization of spatial relations. In C. Freksa, C. Habel, and K. F. Wender, editors, *Spatial cognition. An interdisciplinary approach to representing and processing spatial knowledge*, volume 1404 of *Lecture Notes in Artificial Intelligence*, pages 203–222. Springer-Verlag, Berlin, Heidelberg, New York, 1998.

163. Hanno Walischewski. Learning regions of interest in postal automation. In *Proceedings of the Fifth International Conference on Document Analysis and Recognition*, Bangalore, India, September 1999.

164. Christoph Walther. Mathematical induction. In D. M. Gabbay, C. J. Hogger, and J. A. Robinson, editors, *Handbook of Logic in Artificial Intelligence and Logic Programming – Vol. 2: Deduction Methodologies*, pages 127–228. Oxford, Clarendon Press, 1994.
165. Daniel S. Weld and Johan de Kleer, editors. *Readings in Qualitative Reasoning about Physical Systems*. Morgan Kaufmann, San Mateo, CA, 1989.
166. Alfred N. Whitehead. *Process and Reality*. The MacMillan Company, New York, 1929.
167. Colin P. Williams and Tad Hogg. Exploiting the deep structure of constraint problems. *Artificial Intelligence*, 70:73–117, 1994.

Index

Lecture Notes in Artificial Intelligence (LNAI)

Lecture Notes in Computer Science